Health Information Science

With the development of database systems and networking technologies, Hospital Information Management Systems (HIMS) and web-based clinical or medical systems (such as the Medical Director, a generic GP clinical system) are widely used in health and clinical practices. Healthcare and medical service are more data-intensive and evidence-based since electronic health records are now used to track individuals' and communities' health information. These highlights substantially motivate and advance the emergence and the progress of health informatics research and practice. Health Informatics continues to gain interest from both academia and health industries. The significant initiatives of using information, knowledge and communication technologies in health industries ensures patient safety, improve population health and facilitate the delivery of government healthcare services. Books in the series will reflect technology's cross-disciplinary research in IT and health/medical science to assist in disease diagnoses, treatment, prediction and monitoring through the modeling, design, development, visualization, integration and management of health related information. These technologies include information systems, web technologies, data mining, image processing, user interaction and interfaces, sensors and wireless networking, and are applicable to a wide range of health-related information such as medical data, biomedical data, bioinformatics data, and public health data.

More information about this series at http://www.springer.com/series/11944

Jiming Liu • Shang Xia

Computational Epidemiology

From Disease Transmission Modeling to Vaccination Decision Making

Springer

Jiming Liu
Department of Computer Science
Hong Kong Baptist University
Kowloon, Hong Kong

Shang Xia
Department of Computer Science
Hong Kong Baptist University
Kowloon, Hong Kong

ISSN 2366-0988 ISSN 2366-0996 (electronic)
Health Information Science
ISBN 978-3-030-52108-0 ISBN 978-3-030-52109-7 (eBook)
https://doi.org/10.1007/978-3-030-52109-7

This Springer imprint is published by the registered company Springer Nature Switzerland AG
The registered company address is: Gewerbestrasse 11, 6330 Cham, Switzerland

This book is dedicated to all the people around the world, who came together to fight against the novel coronavirus (COVID-19) pandemic.

Jiming Liu and Shang Xia

Preface

What Can We Learn from COVID-19?

To see a World in a Grain of Sand
And a Heaven in a Wild Flower
Hold Infinity in the palm of your hand
And Eternity in an hour

William Blake (1757–1827) *Auguries of Innocence*

The title of this preface may look a bit unusual for a research monograph. Nevertheless, this was indeed the kind of questions that came to our mind when we were writing it.

This book was born in a special time. As it was being written and published, the world was enduring one of its greatest challenges in decades, if not in centuries. The novel coronavirus, known as COVID-19, had rapidly spread to around 200 countries and territories in 6 continents (with only Antarctica untouched) within a few months, resulting in more than 5 million people infected and over 300,000 deaths (as of May 2020). All the people, no matter where they were and who they were, found themselves caught right amid this most unprecedented global crisis, with devastating casualties, country lockdowns, service/business shutdowns, and possible economic meltdown.

The world is truly in a state of emergency, a time of great uncertainty and anxiety.

Yet, as in the history of human civilization, we all should be hopeful that humankind will be able to learn and prevail in the end. There will be no exception this time. One of the important lessons that we can probably learn from the Mother Nature in this global fight against COVID-19 is that **only by being united as one, as humanity, working together to remove the barriers of races,**

nation-states, political ideologies, religions, and special interests, and coexisting harmoniously in an increasingly interconnected and interdependent world, can human beings be saved. This also calls upon scientists to rethink their roles and social responsibilities, to rediscover the world, and to advance sciences beyond the usual disciplinary boundaries. Under such a unique circumstance, the theme of this book becomes particularly appropriate, as it attempts to show how disciplines such as computer science, systems science, and epidemiology can converge and address some of the most pressing, socially relevant issues in eradicating diseases.

The contents presented in this book reflect part of our ongoing initiatives at Hong Kong Baptist University (HKBU), which are aimed to address several important problems in infectious disease epidemiology and to solve them in a systematic way through the developed computational models, methods, tools, and case studies. Some examples of the problems are as follows:

- How has the field of epidemiology evolved (Chap. 1)? How can data-centric technologies be incorporated? (Chaps. 1 and 7)
- How can the heterogeneous nature of disease transmission be modeled and characterized? (Chap. 2)
- How can we strategically plan and achieve disease interventions (Chap. 3)?
- How can we take into consideration the human (individual and social) aspects of decision-making in disease interventions? (Chaps. 4–6)
- How can the epidemiological challenges be best addressed from a systems perspective? (Chap. 7)
- What promises does systems epidemiology hold? What is the best way to pursue it? (Chap. 7)

Solutions to the above problems can help governments, public health policy-makers, scientists, and front-line practitioners in seeing the current and future global health challenges, such as COVID-19, from a systematic, data-driven computational modeling perspective, and hence developing the corresponding effective intervention strategies. For instance, the solutions provided in this book can help respond to the following questions in the case of COVID-19: Once a coronavirus vaccine becomes available, what will be the best (scientifically sound and yet practically acceptable) way to administer the limited supplies? Who will have the priorities? Will there be enough people to take the vaccine, so that the target coverage (herd immunity) can be achieved? How will people make their vaccination decisions?

The book is intended to serve as a reference book for researchers and practitioners in the fields of computer science and epidemiology, who may read Chaps. 1 and 7 of the book first, to gain a holistic view of the domain, prior to reading Chaps. 2–6 for further studies on the specific problems and issues involved.

Together with the provided references for the key concepts, methods, and examples being introduced, the book can readily be adopted as an introductory text for undergraduate and graduate courses in computational epidemiology as well as systems epidemiology and as training materials for practitioners and field workers, who may study the book in the regular order of Chaps. 1–7 and then revisit Chaps. 2–6 to extend some of the topics and problems.

Hong Kong
Hong Kong
May 2020

Jiming Liu
Shang Xia

Acknowledgements

Jiming Liu is extremely grateful for the rigorous foundation development in Physics that he acquired from East China Normal University in Shanghai in the late 1970s and early 1980s and for the mind-opening education and enriched inquiries in Philosophy, Cybernetics, and Psychology that he gained through the most inspirational teachings of David Mitchell, Gary Boyd, and Gordon Past, as well as other thinkers and visionaries, from Concordia University in Montreal in the mid-1980s. Both periods have profoundly impacted him throughout his career and life. He would like to acknowledge the amazing collegiality and friendships that he has enjoyed in more than three decades from many of his mentors, colleagues, collaborators, and students in Montreal before 1994, in Windsor in 2006–2007, and in Hong Kong since 1994, who have not only accompanied, but also enlightened, him throughout his odyssey of intellectual discovery, exploration, and wonder. For the past ten years, he has made special efforts in developing solutions to address real-world problems, such as global health and infectious disease epidemiology in particular, from the novel perspectives of complex systems, network science, machine learning, and autonomy-oriented computing. For this and other rewarding journeys, he would like to express his heartfelt gratitude to: Xiao-Nong Zhou (as well as dedicated colleagues) of National Institute of Parasitic Diseases (NIPD) at Chinese Center for Disease Control and Prevention (China CDC), with whom he co-established the Joint Research Laboratory for Intelligent Disease Surveillance and Control; his long-time colleagues as well as collaborators and supporters at Hong Kong Baptist University (HKBU), William Cheung, Pong Chi Yuen, Yiu-ming Cheung, Yang Liu, among so many others; his previous postdoctoral fellows and research collaborators, Bo Yang, Zhiwen Yu, Xiaofeng Xie, Qing Cai, Zhanwei Du, etc.; his earlier research students, Shang Xia, Benyun Shi, Chao Gao, Li Tao, Xiaolong Jin, Hongbing Pei, Hechang Chen, Xiaofei Yang, Shiwu Zhang, Hongjun Qiu, Jianbing Wu, Qi Tan, Jinfu Ren, and many more. Also, he would like to thank HKBU as a whole for the trust and opportunities to shape and contribute to the university environment in the capacities of Chair Professor in Computer Science, Head of Computer Science Department, Associate Dean (Research) of Faculty of Science, Dean of Faculty of Science, and Associate Vice-President (Research) of the university, to make it the most conducive place for scholarship. He would like to thank Hong Kong Research Grants Council (RGC) for the funding support over the years; a number of grants have been awarded to specifically support his team's research on understanding and solving epidemiological problems through the exciting routes of computer science, machine learning, and artificial intelligence. Last but the foremost, he would like to express his deepest thanks to his wife M.L. and his daughters I.Y.Y. and B.Y.X. for their long-lasting love and the most wonderful time.

Shang Xia would like to express his sincere gratitude to Prof. Jiming Liu for his enlightening, patience, motivation, enthusiasm, and profound knowledge. Without his encouragement and persistence, this book could not be accomplished. He would like to express his sincere gratefulness to Computer Science Department at Hong Kong Baptist University (HKBU), where he acquired his PhD degree, benefited a lot from the most inspirational guidance, and enjoyed a fulfilling campus life. For this rewarding journey in Hong Kong, he would sincerely express his heartfelt gratitude to Dr. Benyun Shi, Dr. Li Tao, and Dr. Yang Liu, from whom he benefited their collaboration and support. The sincere thanks also go to Prof. Xiao-Nong Zhou and the National Institute of Parasitic Diseases at Chinese Center for Disease Control and Prevention for the great support for his academic career and research. Last but not least, he would like to thank his family: his wife Yao Q.Q. and his daughters Yoyo and Xiuxiu for their caring, love, and support in this wonderful life.

Both authors wish to express their special thanks to Dr. Yang Liu for his great efforts in proofreading the manuscript and offering excellent editorial suggestions and help.

Hong Kong
Hong Kong
May 2020

Jiming Liu
Shang Xia

Contents

Abbreviations

ACIP	Advisory Committee on Immunization Policy
AEFI	Adverse events following immunization
BPA	Basic probability assignment
CHP	Centre for Health Protection
DST	Dempster–Shafer theory
H1N1	Influenza A virus (H1N1)
H7N9	Avian influenza A virus (H7N9)
HSI	Human swine influenza
HSIVP	Human Swine Influenza Vaccination Programme
MMR	Measles–mumps–rubella
NVAC	National Vaccine Advisory Committee
SARS	Severe acute respiratory syndrome
SEIR	Susceptible–exposed–infectious–recovered
SIR	Susceptible–infectious–recovered
SIS	Susceptible–infectious–susceptible
SIT	Social impact theory
STD	Sexually transmitted disease

Notation

S	Population in susceptible compartment
I	Population in infectious compartment
R	Population in recovered compartment
N	Overall population
S_i	Susceptible subpopulation i
I_i	Infectious subpopulation i
R_i	Recovered subpopulation i
N_i	Overall subpopulation i
α	Infectivity
β	Susceptibility
λ	Infection rate
μ	Transmission rate
γ	Recovery rate
c_{ij}	Contact frequency between two subpopulations i and j
R_0	Basic reproduction number
R_t	Effective reproduction number
$\mathbf{C}^H$	Contact matrix for household setting
$\mathbf{C}^S$	Contact matrix for school setting
$\mathbf{C}^W$	Contact matrix for workplace setting
$\mathbf{C}^G$	Contact matrix for general community setting
$\mathbf{C}$	Contact matrix for overall social setting
Φ	Social settings (H, S, W, G)
r^H	Household contact coefficient
r^S	School contact coefficient
r^W	Workplace contact coefficient
r^G	General community contact coefficient
$\mathbf{K}$	Disease reproduction matrix or next-generation matrix
$\mathbf{A}$	Infectivity matrix, $diag\,(\alpha_1, \ldots, \alpha_N)$
$\mathbf{B}$	Susceptibility matrix, $diag\,(\beta_1, \ldots, \beta_N)$
$\mathbf{S}$	Susceptible population matrix, $diag\,(S_1, \ldots, S_N)$
$\mathbf{I}$	Infectious population vector, $[I_1 \ldots I_N]^T$

$\rho(\mathbf{K})$	Top eigenvalue of **K**
$\boldsymbol{x}_1$	Top left eigenvector of **K**
$\boldsymbol{y}_1$	Top right eigenvector of **K**
N_i^{vac}	Number of vaccinated neighbors
N_i^{non}	Number of unvaccinated neighbors
w_{ij}	Social closeness between two connected individuals i and j
$\hat{\lambda}_i$	Perceived infection rate for subpopulation i
$\hat{\beta}$	Perceived susceptibility
θ	Herd immunity threshold
ζ	Cost of disease infection
ξ	Cost of vaccination
r_c	Cost ratio $r_c = \xi/\zeta$
σ_i	Vaccination decision
$\hat{\sigma}_i$	Cost-minimized choice
$\tilde{\sigma}_i$	Social opinion from connected neighbors
ι_i^{vac}	Social influence for vaccination
ι_i^{non}	Social influence against vaccination
$\triangle\iota_i$	Influence discrepancy
ν	Responsiveness to influence discrepancy in Fermi function
$P(\triangle\iota_i)$	Probability generated from Fermi function
r_f	Conformity rate
G	Social network, $G = \langle V,\ L\rangle$
V	Network nodes (individuals)
L	Network links (interactions)
Θ	Universal set of vaccination decision, {Yes, No}
ϕ	Empty set
2^{Θ}	Power set, $\{\phi,\ \{Yes\},\ \{No\},\ \Theta\}$
$m(\cdot)$	Basic probability assignment
$m(Yes)$	Belief value of vaccination
$m(No)$	Belief value of non-vaccination
$m(\Theta)$	Belief value of no decision (uncertainty)
m_i	Set of belief values
m_i'	Updated belief values
m_i^e	Obtained awareness about negative events
m_{dis}^e	Belief values generated by a severe disease infection
m_{vac}^e	Belief values generated by a vaccine adverse event
f	Coefficient of awareness fading
ϵ	Reporting rate of severe disease infections
κ	Reporting rate of vaccine adverse effects

Chapter 1
Paradigms in Epidemiology

Epidemiology deals with "the study of the occurrence and distribution of health-related states or events in specified populations, including the study of the determinants influencing such states, and the application of this knowledge to control the health problems" [1]. As defined by MacMahon et al. [2], epidemiology is interdisciplinary by nature, concerning the sciences of etiology, genetics, biology, pharmacy, geography, ecology, as well sociology and human behavior. Epidemiological studies motivated by combating infectious diseases mainly focus on four aspects of challenges, as follows: (1) pattern analysis, by investigating the spatio-temporal distributions of the observed disease occurrences; (2) causal inference, by identifying and evaluating associated impact factors; (3) forecasting and prediction, by evaluating the dynamics of infectious diseases with reference to different scenarios; and (4) policy analytics, by exploring and conducting effective intervention measures.

Toward these ends, the pioneers in epidemiology have provided much useful knowledge to guide efforts in infectious disease control. As pointed out by Merrill [3], epidemiology has evolved from supernatural practices to research based on scientific foundations, from ad hoc reports to systematic investigations of public health events and problems, from ignorance of the causes of diseases to a scientific understanding of their hidden factors, determinants, and outcomes, and from lacking feasible means for solving public health problems to having effective approaches to disease intervention.

Developmental milestones in infectious disease epidemiology can be dated back to the work of Hippocrates (460–377 BC), who examined the influence of environments and attempted to explain how diseases transmit and cause infection in a group of host individuals [3]. Other early studies include the work of John Graunt (1620–1674), who described disease mortality rates by applying statistical and census methods [4], and Thomas Sydenham (1624–1689), who studied disease distribution patterns, moving from an observational to an analytical perspective [5]. In the nineteenth century, John Snow (1813–1858) traced the sources of disease

J. Liu, S. Xia, *Computational Epidemiology*, Health Information Science,
https://doi.org/10.1007/978-3-030-52109-7_1

outbreaks (e.g., cholera in Soho, London, in 1854) and thereafter pointed out the associations of disease outbreaks with social and natural environments [6]. To more formally describe the dynamics of disease transmission, Ross (in 1911) and MacDonald (in 1957) developed a set of mathematical equations and proposed a threshold indicator, named the basic reproduction number, to quantitatively characterize the extent of disease transmission [7].

1.1 Methodological Paradigms

Various methodologies have been developed to address a wide range of challenges in infectious disease control and prevention, and these methods have been applied in epidemiological studies in the past several decades. As stated by Zadoks [8], based on the observation of disease occurrences, *descriptive methods*, such as clustering and hot spot analysis, have been used to analyze the patterns of infectious diseases in terms of their temporal, spatial, and demographic distributions in a population, i.e., to answer the questions of when, where, and who. *Statistical methods*, such as regression or Bayesian inference, can be used to further explore the causal relationships between disease occurrence and the possible impact factors, i.e., to answer the questions of why and how. *Predictive methods*, such as mathematical modeling or computer-based simulation, have been developed to forecast the dynamics of infectious diseases during an epidemic, and identify the most suitable indicators for representing such a dynamic process. Based on these, *prescriptive methods*, such as optimization, or scenario and sensitivity analyses, can be used by public health authorities to decide how to implement the most effective intervention strategies, such as the allocation of pharmaceutical resources (e.g., vaccines and antivirals) and social distancing (e.g., segregation and school closures).

Infectious disease epidemiology has undergone a number of methodological paradigm shifts throughout its development, as highlighted in Fig. 1.1. The typical methods mentioned in the preceding paragraph, i.e., descriptive, predictive, and prescriptive methods, correspond to three of those paradigms (the fourth is introduced in the next subsection and discussed in detail in the final chapter of this book). These three paradigms are (1) *empirical* investigation, (2) *theoretical* modeling, and (3) *computational* modeling. Accordingly, we refer to the epidemiological methodologies based on these paradigms as (1) empirical epidemiology, (2) theoretical epidemiology, and (3) computational epidemiology, respectively.

- **Empirical Methods**
  The paradigm of empirical observation and investigation is well suited to the early stage of epidemiological studies. As mentioned by Rothman et al. [9], it typically involves (1) collecting observational data about disease transmission, i.e., when, where, and who, and associated impact factors, e.g., the characteristics of disease pathogens and host individuals at the microscopic scale, and of etiological and meteorological environments at the macroscopic scale; (2) qualitatively describing or quantitatively analyzing observational data to establish associative

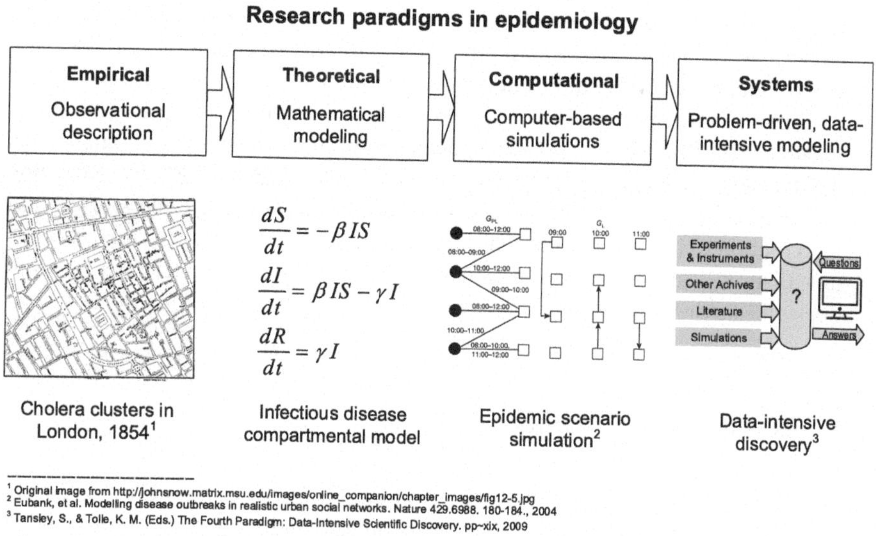

Fig. 1.1 Major methodological paradigms in infectious disease epidemiology

or causal relationships between impact factors and disease transmission; and (3) conducting further experiments or field investigations to test epidemiological hypotheses, usually relating the proposed causes to the observed effects, the findings of which may serve as the foundation for planning and implementing disease intervention.

- **Theoretical Methods**
  The theoretical paradigm in epidemiological studies involves the use of mathematical tools, and is focused on generalizing and characterizing the processes of disease transmission and their interrelationships with various impact factors [10]. Mathematical equations or models are typically constructed to quantitatively describe the dynamics of disease transmission and estimate possible outcomes. By evaluating different conditions under which the models reach convergent, stable, or equilibrium states, public health authorities can potentially make long-term projections and informed decisions on disease intervention. Theoretical epidemiology sometimes draws on certain assumptions and simplifications about the real processes of disease transmission. Meanwhile, it may also require mathematical operations to derive model constructs of the behaviors of various diseases, and use these to infer the disease dynamics and the corresponding intervention measures.

- **Computational Methods**
  With the developments in artificial intelligence, machine learning, data analytics, data mining, and geographic science and information systems, the computational paradigm has rapidly emerged in epidemiological studies. Computational methods are aimed to better characterize and understand the real processes of disease

transmission, by modeling and analyzing the patterns of transmission and quantitatively evaluating the potential outcomes of disease intervention [11]. Primary computational tools that are used comprise computational modeling, simulation, prediction, and optimization, as well as data analytics and visualization, to make the results accessible to public health authorities and epidemiologists. This has further expanded the scope and capabilities of epidemiology for analyzing and predicting the dynamics of disease transmission and the effects of disease intervention in a given population. In addition, public health authorities are now able to more effectively conduct scenario analysis, which facilitates their strategic decision-making.

1.2 Recent Developments

The above-mentioned methods have been in vogue for several decades and have been used to make great contributions to our understanding and ability to combat infectious diseases. However, there remain a number of challenges. As schematically illustrated in Fig. 1.2, these challenges come from emerging and re-emerging infectious diseases, which are significantly correlated with multiple impact factors and their interacting effects, such as genetic mutation of disease pathogens/parasites [12], human socio-economic and behavioral changes [13], and environmental and ecological conditions [14, 15].

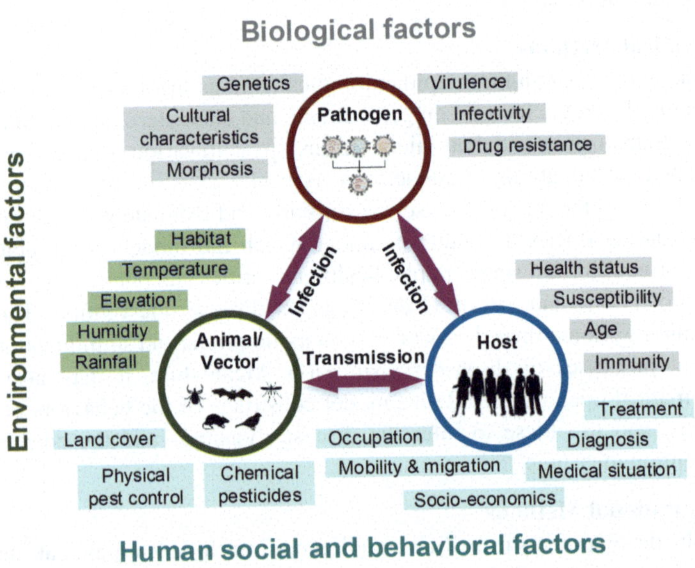

Fig. 1.2 Some interacting components (in circles) and their associated impact factors that can affect the transmission of infectious diseases

Now we examine influenza as an example. It has been shown that a wide range of factors are involved in the dynamic processes of these outbreaks [16], which may include the following: (1) pathogenic factors, such as viral genetic recombination and the expression of pathogens; (2) host factors, such as the immunity of people at different ages; (3) social and behavioral factors, such as people's movement or travel activities; and (4) policy factors, such as disease intervention measures. Furthermore, these factors closely interact with each other. For instance, disease pathogens are carried by humans as they travel, which accelerates the recombination of different types of viruses. However, the implementation of disease intervention, e.g., school closures, changes people's contact behaviors, effectively cutting off the route of disease transmission.

This highlights how various factors can interrelate and interact at various scales. Crucially, the coupling and interactive relationships among those impact factors can determine the intrinsic (yet possibly hidden) spatial, temporal, and social mechanisms of disease transmission. These mechanisms can involve systemic characteristics, such as feedback, saturation, bifurcation, and chaos, thus posing new challenges for comprehensive epidemiological investigations [17].

Effective intervention measures rely on biomedical understanding of disease pathogens/parasites, descriptive studies of spatio-temporal patterns of disease occurrences, and causal analysis of impact factors. In addition, predictive explorations of the trends of disease transmission, i.e., the mechanistic interactions among the components of the transmission process, are also key to understanding and combating infectious diseases. For example, an early warning system for an emerging infectious disease, like COVID-19, requires knowledge about the possible geographic routes of disease transmission, such as human air-travel networks [18]. The prevention of zoonotic and vector-borne diseases, like COVID-19 and malaria, requires both environmental and ecological changes of animal/vector species to be addressed [19], as well as human migrant and mobile behaviors [20]. Furthermore, the effectiveness of disease intervention measures depends on the efficacy of resource allocation, compliance of targeted host populations, and responsive feedback to environmental modifications.

In addition to the above-mentioned challenges, epidemiological studies also face new opportunities in the present and future data-centric era, enabled by the confluence of data from various sources and the development of modeling and analytical tools in data science [21]. For example, a global disease surveillance system connects the health agencies of its member countries and partners at different levels, via local, regional, national, and international organizations [22]. This surveillance system can be used for managing and sharing historical records and reports on when and where specific people have been infected by certain kinds of diseases.

Other data sources are also helpful for analyzing and modeling potential disease transmission. For example, remote sensing data from satellites can readily be utilized for mapping the meteorological and ecological conditions of local or global environments [23, 24].

Another important source of data is Internet-based media, which can serve as an informative channel for revealing individuals' health-related behaviors and opinions. For example, Google Flu Trends was earlier used to assess the transmission of influenza virus [25], and the use of Internet search data was demonstrated to be effective in predicting dengue fever [26].

In view of these challenges and opportunities, it is imperative that new methodologies and paradigms are developed that offer novel perspectives and methods for comprehensive investigation of disease dynamics and associated impact factors, thus expanding our capabilities to understand, predict, control, and prevent the transmission of infectious diseases.

1.3 Infectious Diseases and Vaccination

Faced with the threat of infectious diseases, implementing timely and effective disease intervention measures is critical for preventing mortality and debilitating morbidity, and reducing the socio-economic losses. Various types of intervention measures have been widely studied and adopted for these purposes. For example, immediate isolation/quarantine can prevent transmission during an influenza-like epidemic [27, 28]. The mass prophylactic use of antiviral drugs can reduce the vulnerability of susceptible individuals exposed to infectious diseases [29]. Interventions by social distancing (e.g., school closures and workplace shutdowns) can lower the frequency of contacts among the host population and, hence, reduce the probability of transmitting diseases between susceptible and infectious individuals [30, 31].

Besides the above-mentioned intervention measures, vaccination has been regarded as one of the most effective methods for preventing infectious diseases, due to the effect of vaccine-induced herd immunity (i.e., immunizing a certain portion of the host population provides indirect protection for the unimmunized individuals [32]). That is to say, to prevent a potential outbreak, the vaccination coverage in a host population needs to be above a critical level for inducing the effect of herd immunity, known as herd immunity threshold. In practice, it remains a continual challenge for public health authorities to achieve such a threshold of vaccination coverage for preventing disease outbreaks.

The task is challenging due to a series of reasons. For one thing, although significant progress has been made over the years in vaccine development, the capacity for providing adequate and timely vaccine doses remains a concern, especially when encountering emerging infectious diseases, e.g., 2009 influenza A (H1N1) [33]. Supply restrictions can arise due to many factors, including the time needed for finalizing vaccine compositions, to respond to the constantly evolution of new disease strains [34], the limited capacity for vaccine manufacturing and logistics [35], and the difficulties in access to and uptake of vaccines due to poor delivery infrastructures and economic constraints, especially in developing countries [36]. In such situations, public health authorities in charge of vaccination programs face the question of how to allocate a finite number of available vaccine doses to most effectively prevent disease transmission. For example, the World Health Organization (WHO) has strongly suggested that each country should respond to a possible shortage of vaccine supplies by deciding in advance which groups should have access [37].

Furthermore, the public acceptance of a vaccination program will crucially affect the actual level of vaccine uptake: any loss of confidence in vaccine safety and efficacy will lead to huge gaps between the level of public vaccination willingness and the level needed to contain disease transmission. Historically, societies have experienced several events of vaccine refusal, e.g., the pertussis vaccine scare in the 1970s [38], the decline of measles-mumps-rubella (MMR) vaccine uptake in the 1990s [39, 40], and the rise and popularity of anti-vaccination movements [41, 42]. The rejection of vaccination and the subsequent decline of vaccine uptake have brought about outbreaks of certain vaccine-preventable diseases that were thought to no longer be threats to humankind [43, 44].

In view of this, an in-depth understanding of individuals' voluntary vaccination compliance is urgently required. It has been found that public acceptance of vaccination, which amounts to individuals' decisions on whether or not to take vaccines, are affected by a mixture of cultural, behavioral, and socio-economic factors. For example, the public may have doubts about vaccine safety and efficacy due to scare stories around the adverse effects of vaccination [45, 46]. Or, behaving in their own self-interest, individuals may be inclined not to get vaccinated if enough other people have been vaccinated [47, 48]. The affordability and convenient accessibility of new vaccines are also of importance for individuals considering vaccination, especially in developing countries [49, 50].

Furthermore, the rapid emergence of online social media, e.g., Facebook and Twitter, allows opinions, whether for or against vaccination, to spread broadly and immediately among the population [51]. Therefore, social influences play an increasingly important role in individuals' vaccination decisions. In this regard, an individual's decision on whether or not to vaccinate himself/herself is no longer a personal affair, but will affect the decisions of others, and collectively determine the final coverage of a vaccination program.

Clearly, there exists an urgent need for more systematic studies of vaccination at both population and individual levels, and thereby improve the efficacy of vaccination programs for preventing the outbreak of infectious diseases.

1.4 Objectives and Tasks

In this book, we examine the dynamics of disease transmission in a host population, in which individuals' contact relationships are inferred from the socio-demographic data. Based on such a description of disease transmission, we address the problem of vaccine allocation by developing a novel prioritization method that targets certain subpopulations to most effectively reduce disease transmission. Furthermore, to investigate individuals' acceptance of vaccination, we present decision models to characterize individuals' voluntary vaccination and evaluate the impact of social influences and individuals' subjective perception on the effectiveness of disease intervention by vaccination.

1.4.1 Modeling Infectious Disease Dynamics

As aforementioned, the dynamics of disease transmission depend on many disease- and host population-related factors. To characterize the heterogeneity of a host population, we need to consider its age structure, and then construct a compartmental model to describe disease dynamics with respect to individuals' age-specific variations, i.e., the heterogeneity in terms of age-specific infectivity and susceptibility, as well as cross-age contact relationships, as in the case of COVID-19 transmission [52].

For the purpose of demonstration in this book, we consider the real-world scenario of the 2009 Hong Kong H1N1 influenza epidemic, and calibrate our demonstration parameters with reference to the epidemiological characteristics of influenza A (H1N1). As detailed information about the actual contacts among age-specific subpopulations is often unavailable, we exploit a computational method to infer the contact relationships in terms of individuals' cross-age contact frequencies from the census data in Hong Kong. Specifically, we represent individuals' actual contacts as cross-age contact frequencies within four specific social settings, i.e., school, household, workplace, and general community. We then estimate the overall contacts that account for disease transmission by incorporating four setting-specific contact frequency matrices, which are weighted with the coefficients corresponding to the proportions of individuals' contacts within the considered social settings.

To computationally evaluate our model, we carry out a series of simulation-based experiments to examine its predicted disease dynamics. That is, we validate our epidemic model by comparing the model predictions with the real-world observations, in terms of the daily new infection cases and the age-specific attack rates, i.e., the proportion of infected individuals in each subpopulation. In essence, we reproduce the dynamics of disease transmission based on the heterogeneity of the age-structured host population. The results, as we describe later, can serve as the basis for further discussions on vaccine allocation methods and on individuals' voluntary vaccination.

1.4.2 Modeling Vaccine Allocation Strategies

The heterogeneity of the host population means that the disease-preventing effects of vaccination in individuals of different ages can vary markedly. An immediately related practical question is how to allocate a finite number of vaccine doses to most effectively reduce disease transmission; crucially, this requires knowledge of the effectiveness of this intervention. In this book, we focus on developing a problem-solving method for answering this question. Specifically, we develop a computational method for identifying the relative priority of each subpopulation, by evaluating the effectiveness of age-specific vaccination in reducing disease transmission. We examine the effects of disease intervention on containing disease transmission by measuring the reproduction number corresponding to the age-specific heterogeneity of a host population. By doing so, we identify subpopulations whose vaccination will lead to the greatest reduction in disease transmission, by considering the marginal effects of reducing the reproduction number in cases of age-specific vaccine allocation.

Unlike the existing optimization-based methods, this proposed vaccine allocation method has the following distinct characteristics:

- The method utilizes prior knowledge about individuals' age-specific susceptibility and infectivity, real-time disease prevalence in each subpopulation, and the basic patterns of individuals' cross-age contact frequencies within each social setting. Moreover, it does not rely on detailed information about individuals' actual contacts, nor the potential changes in these contacts in response to disease transmission, which would be difficult to rapidly and accurately determine in practice.
- The method is designed to most effectively reduce disease transmission by allocating a finite number of vaccine doses to certain target subpopulations. The identified vaccination priorities can be adaptively regulated based on the dynamics of disease transmission, i.e., the number of vaccine doses suggested to be allocated to each subpopulation can be dynamically adjusted according to the latest progress of disease transmission and vaccine supply.
- The method incorporates the effects of other disease intervention measures being implemented simultaneously with vaccination, e.g., individuals' contact reduction. Therefore, in the situation of integrated disease intervention, the method can provide more accurate and effective solutions for vaccine allocation.

We apply above-mentioned method to the real-world scenario of the 2009 Hong Kong H1N1 influenza epidemic to identify the relative priorities of subpopulations for disease intervention in Hong Kong. The results show that this method of prioritizing age-specific subpopulations for vaccine allocation and social settings for contact reduction can readily improve the effectiveness of disease transmission-containing efforts.

1.4.3 Modeling Vaccination Decision-Making

In a voluntary vaccination program, individuals' decisions on whether or not to uptake vaccine crucially affect the level of vaccination coverage and, thus, the effectiveness of disease intervention. In this regard, modeling and evaluating individuals' vaccination decision-making would provide useful information for public health authorities on how to improve the effectiveness of vaccination programs [53].

Researchers have typically utilized payoff-based approaches to characterize individuals' vaccination decision-making with respect to the perceived risks and benefits of vaccination. Moving beyond that, we consider the fact that whether an individual does or does not get vaccinated is also influenced by the decisions of others, i.e., social influences. We thus view individuals' voluntary vaccination as an integrated decision-making process that incorporates both a cost analysis of vaccination and the impact of social influences.

Our integrated decision model is an improvement over the existing models, and has several interesting features, as follows:

- We model an individual's vaccination decision-making as an integrated process that balances his/her self-initiated cost minimization and the social influences of others' decisions. Moreover, this model introduces a parameter, called the conformity rate, to modulate individuals' tendency toward two decision-making mechanisms: an individual will adopt his/her own cost-minimized decision, or the social opinion of his/her interconnected neighbors.
- Based on the existing studies in which the social influences on the process of opinion formation have been addressed, we further consider the heterogeneity of individuals' social relationships, i.e., how individuals are socially interconnected. Computationally, we model and characterize the effect of networked social influences on individuals' vaccination decisions based on Social Impact Theory (SIT).
- Based on this new model, we examine the effects of social influences on individuals' decisions and on the effectiveness of disease intervention (vaccination coverage), with respect to three determinants: (1) the relative costs of vaccination and infection; (2) individuals' conformity to social influences, i.e., conformity rate; and (3) individuals' initial level of vaccination willingness.

We parameterize the integrated decision model based on the real-world scenario of the 2009 Hong Kong H1N1 influenza epidemic, and perform a series of simulation-based experiments to infer the coverage of voluntary vaccination programs as a result of individuals' decision-making. The results indicate that individuals' vaccination decision-making can be affected by both the associated costs and their conformity to social influences. Thus, it becomes necessary for public health authorities to estimate the level of individuals' acceptance of vaccination prior to the start of a voluntary vaccination program, as well as to rapidly assess and enhance the effectiveness of their adopted vaccination policies, e.g., providing certain financial subsidies to reduce the cost of vaccination.

1.4.4 Modeling Subjective Perception

It has long been observed that the spread of awareness about an epidemic via social media can affect individuals' opinions and behaviors concerning an epidemic. In the case of an emerging infectious disease, it can be difficult for individuals to become informed about the disease and/or a newly developed vaccine prior to their decision-making. In such a case, the spread of awareness about disease severity and vaccine safety could affect individuals' subjective perception about vaccination and, hence, substantially affect their decisions [54].

To gain a better understanding of individuals' voluntary vaccination, we develop a belief-based decision model to evaluate the effect of the spread of awareness on individuals' decision-making and on the effectiveness of disease intervention. Compared with the existing studies on modeling individual-level vaccination decision-making, this belief-based model has the following unique properties:

- Unlike existing decision models that represent decision-making as a binary problem, we consider the role of uncertainty in individuals' vaccination decision-making. Specifically, the situation in which an individual has made no firm decision can be considered as a state of "yet to decide", due to uncertainty. In this regard, we introduce three belief variables to characterize the possible decision response from an individual, namely that he/she will accept or reject the vaccine, or has not yet decided.
- We further consider the fact that individuals' decisions depend on their subjective perception about whether or not vaccination is acceptable. Moreover, awareness of disease severity and vaccine safety can spread from person to person—akin to a disease itself—and will substantially affect their subjective perception of vaccination.
- To model the spread of awareness, we utilize various real-world online social networks to characterize the structure of individuals' social relationships. Thereafter, we further extend Dempster-Shafer Theory (DST) to computationally model the propagation and evolution of individuals' beliefs, as well as their decision-making, having incorporated the awareness obtained from their socially interconnected neighbors.

We investigate the effect of the spread of awareness on individuals' vaccination decision-making with respect to three considered impact factors, based on a series of simulations of the 2009 Hong Kong H1N1 influenza epidemic. First, the reporting rates of severe infection and adverse effects of vaccination are used to represent the frequencies of these topics, which tend to draw public attention. Next, we consider the coefficient of awareness fading, a parameter used to quantify the information flows among individuals. Finally, we examine the effect of disease reproduction number, which corresponds to the severity of an epidemic.

The simulation results show that the reporting rates will determine the number of vaccinated individuals and the time at which they receive vaccination. A higher fading coefficient will significantly reduce individuals' vaccination willingness.

A larger value of disease reproduction number will enhance the proportion of vaccinated individuals, although this cannot compensate for the growth of the infected population size resulting from a more severe disease outbreak.

1.5 Summary

In this book, we develop a computational modeling approach to evaluate and guide the implementation of different intervention measures for controlling infectious diseases. We focus on the following topics in the remaining chapters:

In Chap. 2, we provide a general description of the concepts and related computational models and tools for characterizing disease transmission dynamics in a heterogeneous host population. Specifically, we introduce the concepts of compartmental modeling for describing disease transmission in an age-structured host population. Then, we present a computational method for inferring the cross-age contact patterns of the population. Finally, we parameterize and validate the epidemic model with a real-world epidemic scenario, which serves as the basis for our further discussions on vaccine allocation and individuals' voluntary vaccination.

In Chap. 3, we develop a prioritization method for identifying target subpopulations for vaccine allocation that would enable us to most effectively reduce disease transmission. We walk through a series of simulation-based experiments that evaluate the performance of such a vaccine allocation method in improving the effectiveness of disease intervention.

In Chap. 4, we examine vaccination decision-making from the perspective of individuals. In particular, we show how to model individuals' decision-making processes in response to a voluntary vaccination program. We use computational modeling to perform a game-theoretic analysis of the costs and benefits of vaccination with respect to individuals' social relationships. Then, we experimentally examine the level of vaccination coverage, based on this game-theoretic model, through a series of simulations of voluntary vaccination.

In Chap. 5, we introduce an extended decision model that additionally addresses the effect of social influences on an individual's decision whether to undergo voluntary vaccination. In the extended decision model, we utilize the SIT to further characterize social influences with respect to individuals' social relationships. We evaluate the effect of social influences by computing the level of vaccination coverage through a series of simulations of voluntary vaccination based on such an integrated decision model.

In Chap. 6, we present a more complete investigation of voluntary vaccination by modeling and examining the effect of the spread of awareness on vaccination decision-making. In doing so, we develop a belief-based decision model, in which individuals' decisions are affected by their subjective perception of vaccination. In this model, we utilize and extend DST to characterize individuals' belief updates and changes in vaccination decisions accordingly. We evaluate the effect of the spread of awareness by carrying out simulation-based experiments to examine the time course of vaccine administration and disease transmission.

In Chap. 7, we offer a fresh outlook on the latest methodological paradigm in infectious disease epidemiology, which is known as systems epidemiology. Specifically, we introduce the fundamental ingredients of systems thinking, which are essential for viewing and addressing complex epidemiological questions holistically. We then provide detailed systems modeling principles and practical steps that can be followed in future systems epidemiological studies.

Finally, under "References" section, we provide a detailed list of references for further reading and research.

Chapter 2
Computational Modeling in a Nutshell

There are many parallels between epidemiological studies and systems studies. That is, we can view the problems of epidemiology from a systems perspective. In systems studies, the goal of modeling is to develop representations or frameworks, in mathematical or computational languages, that are abstracted from and yet allow for characterization of certain real-world observations. In epidemiological studies, the essence of modeling further entails two important aspects: (1) **Problem-driven conceptual modeling**, which translates certain real-world problems in an epidemiological domain into conceptual models in a theoretical or computational domain. As an example, a metapopulation-based compartmental model may be utilized to describe disease dynamics in an age-structured host population. (2) **Data-oriented real-world grounding**, which requires us to discover ways to embody the developed conceptual models, i.e., model parameterization, by obtaining and utilizing real-world data and statistical analysis of the real-world observations. For instance, the effect of age-specific population contact patterns may be inferred from data by utilizing a computational method. Furthermore, we may parameterize and validate the contact-based model with a real-world scenario, such as the 2009 Hong Kong H1N1 influenza epidemic.

In this chapter, we provide an overview of the most essential models for characterizing infectious disease transmission, namely, epidemic models and contact relationship models. By doing so, we show how some of the questions in infectious disease epidemiology can be approached by modeling and characterizing the dynamics of infectious disease transmission [55].

2.1 Modeling Infectious Disease Dynamics

Accurate predictions about the dynamics of disease transmission play an essential role in controlling infectious diseases. In this regard, mathematical modeling and

J. Liu, S. Xia, *Computational Epidemiology*, Health Information Science,
https://doi.org/10.1007/978-3-030-52109-7_2

analysis offers important tools for characterizing disease transmission in a host population and evaluating the effectiveness of intervention measures [56]. In this section, we focus on the mathematical characterization of disease transmission based on the contact relationships of individuals responsible for disease transmission. Specifically, we review (1) basic notions and concepts in disease transmission, and (2) existing models for representing disease dynamics.

The spread of an infectious disease results from the transmission of the contagious pathogen between hosts through their direct or indirect contact. During an epidemic, the dynamics of disease transmission within a host population involve multiple determinants, including disease-dependent biological factors and disease-independent behavioral factors [57]. Disease-dependent biological factors refer to the viral or bacterial features of the pathogen, such as the pathogen's life cycle, its dynamics of replication and development within the hosts, and its virulence and sensitivity to drug treatments. Biological factors also include the physiological conditions of a host population, e.g., individuals' innate or acquired immunity. Disease-independent behavioral factors that affect disease transmission include the demographic and social structures of a host population, e.g., age distribution and the contact relationships among different age groups. The contacts responsible for disease transmission depend on the specific pathogen and the transmission route. Taking influenza-like respiratory diseases as an example, individuals' contacts during informal gatherings, such as face-to-face conversations and handshakes, can result in disease transmission [58].

In infectious disease epidemiology, one important indicator, called the basic reproduction number R_0, has been widely used to evaluate the force of disease transmission at the population level. It is defined as the number of secondary infection cases generated by a typical infected individual during their entire period of infectiousness in a completely susceptible population [59, 60]. R_0 has been used to indicate whether or not an infectious disease will cause an outbreak in a host population. Specifically, if R_0 is greater than one, outbreaks can occur. Conversely, if R_0 is less than one, disease transmission in a host population will naturally disappear. Based on this property, the estimation of R_0 provides a feasible way of predicting disease outbreaks and evaluating the effectiveness of disease control. That is to say, to prevent disease transmission in a host population, the adopted disease interventions should ensure that R_0 is less than the threshold of outbreak.

2.1.1 *Infectious Disease Models*

In studies of infectious diseases, mathematical modeling and analysis of disease transmission is crucial for control and prevention. Infectious disease models are formulated to characterize disease dynamics in terms of how susceptible individuals will potentially be infected when exposed to infectious individuals, hence offering a quantitative description of disease incidence and prevalence within a host population. Therefore, developing infectious disease models helps reveal the

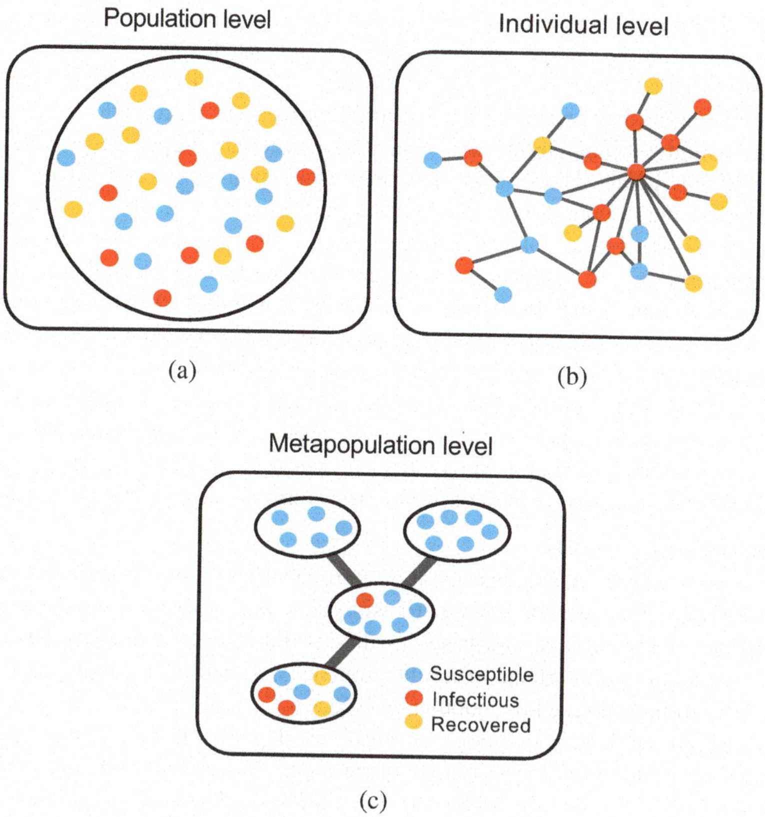

Fig. 2.1 Epidemic models: (**a**) population-level compartmental models, (**b**) individual-level network-based models, and (**c**) metapopulation-level models

mechanisms of disease transmission and evaluate the outcomes of potential control and prevention measures during an epidemic. In what follows, we look at three types of models that have been used to represent disease transmission at different levels of a host population. As shown in Fig. 2.1, the models are (1) population-level, (2) individual-level, and (3) metapopulation-level models.

Population-Level Models

Early epidemic models, known as compartmental models, were developed based on the assumption that individuals randomly have contact with each other (the homo-mixing assumption as illustrated in Fig. 2.1a) [61, 62]. During the time course of disease transmission, each individual belongs to one of the compartments, namely, uninfected and susceptible (S), infected and not yet infectious (E, or exposed), infected and infectious (I), and recovered with immunity (R). Based on the homo-mixing assumption, the dynamics of disease transmission can thus be modeled as the population transitions among those different compartments. For example, the susceptible-infectious-recovered (SIR) model has been widely used for describing disease transmission, such as seasonal influenza and measles [63, 64].

Many variations have been proposed based on the basic framework of compartmental modeling. For example, the susceptible-infectious-susceptible (SIS) model has been used to characterize the situation of secondary/repeated infections in recovered individuals. The susceptible-exposed-infectious-recovered (SEIR) model characterizes the important infections within the period of time when an individual has been infected but is not yet infectious (i.e., in the exposed compartment E). Furthermore, the S, I, and R compartments may be further divided into several subgroups to reflect the demographic structure (e.g., age) of a host population [56]. Thus, the dynamics of different subpopulations can be taken into account by incorporating the transmission parameters for each subpopulation.

Although compartmental models can provide deterministic descriptions of disease dynamics at the population level, the assumption of homogeneous individuals may not adequately capture the reality. In view of this, new approaches for individual-level characterization of disease transmission have been developed.

Individual-Level Models

In attempts to explicitly characterize the heterogeneity of a host population, which can affect the dynamics of disease transmission, new individual-level epidemic models (also known as network-based models as illustrated in Fig. 2.1b) have been developed to study the effects of the heterogeneity of pathogens, individuals, and their interactions within a host population [65].

During an epidemic, disease transmissions depend on individuals' contact relationships. The contact patterns (frequency and structure of contacts in a host population) can profoundly affect the population-level disease dynamics. Contradicting the homo-mixing assumption that individuals uniformly and randomly have contact with each other, each individual in reality has a different number of relationships with contacts to whom he/she could potentially transmit a pathogen, or from whom he/she could become infected with a pathogen. The structure of individuals' contacts in such a heterogeneous host population can readily be described using a network model, in which nodes denote host individuals and links represent the contact relationships among the interconnected nodes [66]. In a contact network, the degree of a node (the number of links connected to an individual) reflects the number of neighbors through which disease transmissions may take place [67].

As mentioned, compartmental models utilize deterministic differential equations to describe the mean dynamics of disease transmission at the population level, but these cannot adequately characterize the effects of heterogeneity within a host population. Network-based models overcome this shortcoming by modeling individual-level disease transmission in a heterogeneous contact network. Here, the transmission is characterized as a stochastic percolation process, in which each individual may with a certain probability be infected by his/her network neighbors [68]. This probability is computed based on the number of contact relationships an individual may have. In other words, in a heterogeneous contact network, the number of contact relationships one has will determine his/her corresponding chance of getting infected and, hence, the route of disease transmission through the network.

Network-based models provide a way of incorporating individual-level heterogeneity. However, they also have limitations. For example, to construct a contact network of disease transmissions, one must know in advance the detailed characteristics of each individual and the disease-associated contact relationships. In practice, it is difficult, if not impossible, to have such prior knowledge, sometimes even for a small group of host individuals.

Thus far, existing studies have addressed the problem of representing individuals' contact relationships primarily through statistical means, by collecting and analyzing empirical data on individuals' contact activities in certain places/regions for a fixed time period [69]. These studies may provide static and empirical descriptions of individuals' contact relationships (who meets whom, when, where, and how often). However, this approach also has limitations, as follows: (1) it is hard to define the type of contacts responsible for disease transmission, because this depends on the type of disease; (2) it is not always feasible to have such an empirical description of contact relationships for a certain host population; and (3) individuals' actual contacts may change during the spread of an infectious disease, invalidating the empirical knowledge basis on which disease transmissions are predicted.

In this regard, the challenge lies in the lack of accurate and reliable descriptions of individuals' contact relationships, without which the accuracy of model predictions cannot be improved. As discussed in the following subsection, one promising alternative is to develop a computational method to infer individuals' contact relationships from the socio-demographic data of the population. Such inferred contact relationships can serve as a good foundation for modeling age-specific disease dynamics and the corresponding intervention strategies.

Metapopulation-Level Models

As aforementioned, population-level compartmental models based on differential equations provide a deterministic description of the dynamics of disease transmission in a host population, and have the advantage of being tractable. However, the assumption that individuals are homogeneously mixed neglects the variations in the underlying disease transmission. In contrast, individual-level models based on contact networks do address the host population's heterogeneity, but the disease dynamics resulting from a stochastic process are often intractable and sensitive to the specific settings for simulations. In addition, it remains challenging to construct realistic contact networks.

As a solution to bridge the gap between population-level compartmental models and individual-level network models, metapopulation-level models have been developed, which modify the conventional compartmental models by further taking into account the structure of a host population (Fig. 2.1c) [70, 71]. These models address the heterogeneity of the host population by subdividing it into several subpopulations according to certain characteristics of the individuals, such as their age, occupation, and geographic location [72]. Once this is done, the infection dynamics of the subpopulations are characterized, based on their distinct transmission characteristics and interrelationships.

There are some strong reasons for modeling the subpopulations in such a manner. For example, in a disease outbreak, children and the elderly may be more vulnerable to infection than younger adults [73]. School-age students may be more inclined to have contact with individuals of similar ages, while adults of working age tend to mostly associate with each other [74]. To address the challenge of effective and timely infectious disease control, we need to investigate the role of each subpopulation in disease transmission, corresponding to their specific heterogeneous characteristics.

2.1.2 *Age-Specific Disease Transmissions*

The basic compartmental model (susceptible-infectious-recovered model, SIR for short) deals with the dynamics of disease transmission within a single epidemic, in which the birth and death rates of a host population are not taken into consideration [56]. The number of individuals in each compartment is described by the following set of ordinary differential equations:

$$\begin{aligned} \frac{dS}{dt} &= -\lambda S \\ \frac{dI}{dt} &= \lambda S - \gamma I \\ \frac{dR}{dt} &= \gamma I \end{aligned} \tag{2.1}$$

where S, I, and R denote the numbers of individuals in the susceptible, infectious, and recovered/immunized compartments, respectively. The parameter λ is the infection rate, which denotes the proportion of the susceptible population that will become infected during the present time step. The parameter γ is the recovery rate, which describes the rate at which the infectious individuals recover and, therefore, become immunized from secondary infection. In detail, λ can be viewed as a composite of three factors:

$$\lambda = \beta \cdot \frac{I}{N} \cdot \alpha \tag{2.2}$$

where α describes the infectivity of infected individuals and β is the susceptibility of uninfected individuals. N is the total size of the host population. In this case, the basic reproduction number R_0 can be computed as follows:

$$R_0 = \frac{\alpha \cdot \beta}{\gamma} \tag{2.3}$$

We now introduce an age-specific compartmental model, in which individuals are divided into n subpopulations with reference to their ages. Each individual in age group i belongs to one of three infection-associated compartments: susceptible (S), infectious (I), and recovered/immunized (R). Correspondingly, the number of individuals in each compartment is denoted by S_i, I_i, and R_i, respectively. We consider disease transmission within a single circulation season, such that the natural birth and death rates of the population are not taken into account. The total number of individuals in age group i, denoted by $N_i = S_i + I_i + R_i$, is static.

The dynamics during the time course of disease transmission is described by the following set of differential equations:

$$
\begin{aligned}
\frac{dS_i}{dt} &= -\lambda_i S_i \\
\frac{dI_i}{dt} &= \lambda_i S_i - \gamma_i I_i \\
\frac{dR_i}{dt} &= \gamma_i I_i
\end{aligned}
\tag{2.4}
$$

where γ_i represents the rate of recovery, corresponding to the duration of disease infection. λ_i is the infection rate, which denotes the probability of being infected for susceptible individuals in age group i. For each time step, λ_i can be calculated as follows:

$$
\lambda_i = \mu \cdot \beta_i \cdot \sum_{j=1}^{n} \left(c_{ij} \cdot \alpha_j \cdot \frac{I_j}{N_j} \right) \tag{2.5}
$$

where c_{ij} describes the contact frequency between individuals in age groups i and j, α_j measures the infectivity for individuals in group j, which is the probability of disease transmission when an infectious individual has contacts with other susceptible individuals, β_i denotes the susceptibility for individuals in group i, which represents the probability of being infected when a susceptible individual is exposed to infectious contacts, and μ is a constant disease transmission rate for all age groups and can be estimated from R_0 in the initial stage of disease transmission.

2.2 Modeling Contact Relationships

To accurately predict the dynamics of disease transmission, it is crucial to know individuals' contact relationships, as these determine the routes of disease transmission within a host population. In what follows, we examine some of the frequently used contact models. For the identified impact factors, the contact relationships among the host population play an important role in determining the dynamics of disease transmission. In practice, it remains a major challenge to accurately and reliably describe individuals' contact relationships during the course of disease transmission. As a result, it is difficult to rapidly and reliably predict disease transmissions. For this reason, efforts have been made to characterize the social contact relationships based on the empirical data of individuals' contact frequencies and durations [75].

In this section, we review some of the existing studies on characterizing individuals' contact patterns for modeling disease transmission within a host population. Specifically, based on different ways of collecting contact-related information, we consider two types of studies, namely (1) empirical methods that collect individuals' actual contact activities, and (2) computational models that infer/simulate individuals' contact behaviors from data.

2.2.1 *Empirical Methods*

Empirical methods involve collecting information about individuals' actual contact relationships to describe individuals' contact patterns and thereby reveal how a disease is transmitted from one person to another. As mentioned by Keeling [65], we can use three types of empirical methods:

- **Infection Tracing**
  Infection tracing is performed to determine the route of disease transmission for each case of infection, i.e., for each infected individual, infection tracing can identify by whom the individual was infected, and the group of persons who have been infected by the individual. By doing so, infection tracing can collect detailed information about the majority of infection cases and, hence, reveal the probable routes of disease transmission. For example, Haydon et al. constructed epidemic trees with respect to the 2001 U.K. foot-and-mouth outbreak [76], while Riley et al. traced cases of the severe acute respiratory syndrome (SARS) epidemic in Hong Kong [77]. Thus, by constructing a tree-like network of disease transmissions, the corresponding spatial and temporal patterns can be accurately depicted. Moreover, the epidemiological parameters, e.g., disease reproduction number R_0, can be directly estimated from the collected infection cases. Therefore, infection tracing is an individual-based method that can describe the network of disease transmissions within a relatively small population, and provides a quick and simple analysis of disease dynamics.
- **Contact Tracing**
  In contrast to infection tracing, which gathers information on the actual links in the chain of disease transmissions, contact tracing focuses on identifying all of the contact relationships for an infected individual. This method emphasizes the potential routes of disease transmission, in terms of a group of individuals who might be infected due to their contacts with infected individuals. In this case, individuals in the network of contacts can be targeted for treatment or quarantine [78]. Contact tracing has been successfully used to depict the network of various disease transmissions, including that of sexually transmitted diseases

(STDs) [79, 80], as well as air-borne diseases [81, 82]. However, although both infection tracing and contact tracing can provide a contact network to represent the potential routes of disease transmission, such a network covers only a subset of the host population, as it is focused on the contact relationships immediately surrounding the infected individuals.

- **Survey-Based Studies**
  Survey-based studies record individuals' contact activities in detail for the entire host population. For example, Mossong et al. collected diaries from individuals in eight European countries that recorded the characteristics of those individuals' daily contacts with reference to their contactees' age and sex, and the location, duration, frequency, and occurrence of physical contacts [83]. Read et al. launched a detailed diary-based survey of contact in terms of a group of adults' casual and close contact encounters [84]. Hens et al. carried out a 2-day population survey in Belgium to mine social contact patterns for epidemic models [85]. Such survey-based studies have provided comprehensive descriptions of individuals' contact patterns, which can be used to improve the accuracy of epidemic models for infectious disease control.

The above-mentioned empirical studies can provide quantitative descriptions of real networks of individuals' contact relationships. However, there are several limitations to epidemic modeling and disease control. First, it is difficult to define the connections of a contact network in a way that generally represents the routes of disease transmission, as the types of contact responsible for infection are disease-dependent; e.g., influenza spreads through the air, while HIV is transmitted via sexual contact. Next, it is time- and labor-intensive (and thus impractical) to obtain detailed information from the entire population with respect to their contact behaviors. Therefore, it is not always feasible to acquire such an empirical characterization of contact patterns within a specific host population. Further, survey-based studies can only provide static and empirical descriptions of individuals' contact relationships. Finally, individuals' potential behavioral changes, in response to both disease transmission and adopted disease interventions, are not taken into account by surveys, although these can significantly affect the dynamics of disease transmission.

2.2.2 Computational Methods

The challenge for characterizing disease dynamics is to obtain sufficient realistic data to represent the contact relationships among a host population. In this section, we discuss how this challenge can be solved by applying computational models that enable the representation and estimation of individuals' contact relationships and correspondingly characterize the effects of these relationships on disease transmission.

- **Contact Networks**
  Contact networks are essentially graphs, where a graph is a collection of nodes, which are joined by a set of connections, called links. Each link denotes a relationship or an interaction between the nodes it joins. Contact networks can be further categorized as undirected or directed, and the directedness of a network is highly epidemiologically relevant, as it indicates the possible paths of propagation of a contagion.
- **Contact Matrix**
  In a contact network of size N (i.e., with N nodes), a compact way to specify all contact relationships is to utilize an adjacency matrix C, in which elements $c_{ij} = 1$ if a link connects nodes i and j, and zero otherwise. Such a matrix C is also known as a contact matrix, and can be used to describe the individuals' contact relationships, such as the occurrence of pairwise contacts or the frequencies of their contacts [52]. Both contact networks and contact matrices are computational characterizations of individuals' contact relationships. Note that contact networks that are undirected or directed correspond to contact matrices that are symmetric or asymmetric, respectively.

Computational methods to infer contact networks from the demographic and social characteristics of a host population are in practice very useful for infectious disease modeling. For example, in studying the contact network of STDs, simulation-generated networks have been used to characterize the patterns of observed contact relationships, e.g., the role of hub individuals with a large number of partners [86, 87]. Potterat et al. described a risk network for individuals with HIV infection in Colorado Springs, U.S.A., by analyzing the community-wide HIV/AIDS contact-tracing records during the 1980s and 1990s, i.e., sexual and drug-injection partners [88].

In modeling outbreaks of air-borne diseases, simulation-generated contact networks have been used to capture the demographic and social characteristics of the considered host population [89]. For example, Halloran et al. simulated the stochastic spread of smallpox in a community of 2000 people, in which each individual was generated from population's age distribution and household size that can be derived from census data [90]. Meyers et al. investigated the dynamics of SARS by generating a contact network based on the social setting in the city of Vancouver [91]. Eubank et al. explored dynamic bipartite graphs to model the physical contact patterns corresponding to individuals' movements between different locations [92, 93].

2.3 Case Study

In this section, we examine a case study of the 2009 Hong Kong H1N1 influenza epidemic. We discuss how to characterize the heterogeneous contact relationships within an age-structured host population. Note that in this case, an empirical description of individuals' contact relationships is not available and, thus, we need a computational method to infer the cross-age contact frequency and structure from the social-demographic data of the population. In addition, we examine how such an age-structured contact characterization helps assess the dynamics of disease spread. In doing so, we use a compartmental model that is parameterized and validated with respect to the real-world epidemic scenario.

2.3.1 Hong Kong H1N1 Influenza Epidemic

In Hong Kong, the first H1N1 influenza (also known as human swine influenza, HSI for short) infection case was an imported case confirmed on May 1, 2009 [94]. The first reported local case (i.e., the first indigenous HSI infection without an epidemiological link to imported patients) was laboratory-confirmed on June 10, 2009. As of September 2010, there were over 36,000 laboratory-confirmed cases of HSI in Hong Kong [95]. Figure 2.2a shows the daily number of newly reported H1N1 infection cases over the period of 200 days since the disease onset in early May 2009, as reported by the Centre for Health Protection (CHP) of Hong Kong [96].

Figure 2.2b further gives the proportions of disease infections in different age groups. In what follows, we present a data-driven computational model for investigating the dynamics of disease transmission among those age-specific subpopulations.

2.3.2 Age-Specific Contact Matrices

In this section, we discuss how to computationally characterize social contacts based on individuals' contacts in various social settings, namely, school, household, workplace, and general community, which capture the patterns of interest or the likelihoods of individuals having contact with each other. From such patterns, we can further investigate ways of reducing social contact-based disease transmission, as discussed in the next chapter, by enforcing changes to individuals' contacts in these social settings.

Here, we consider susceptible individuals being infected through their social contacts with infectious individuals. The number of disease transmissions among different age groups is therefore determined by the frequencies of contacts among them. Individuals' contact relationships can be represented using a contact matrix, and it has been shown that there exist strong diagonal elements (indicating high contact frequencies) among those aged 5–24 years [83]; this pattern reflects that individuals tend to have contact with others of similar ages within places such as schools. At the same time, there also appear parallel secondary diagonals that

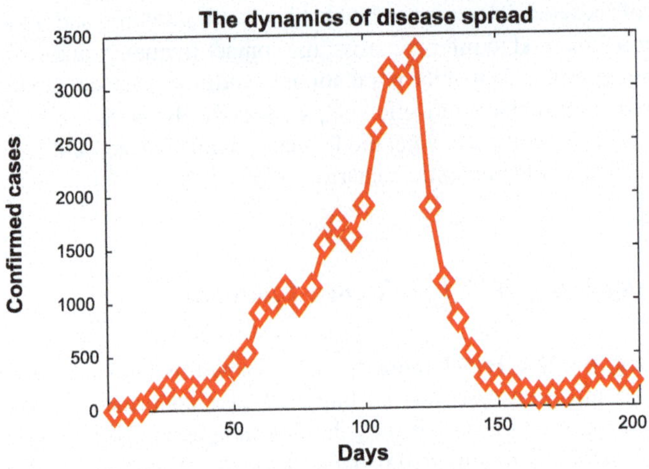

(a)

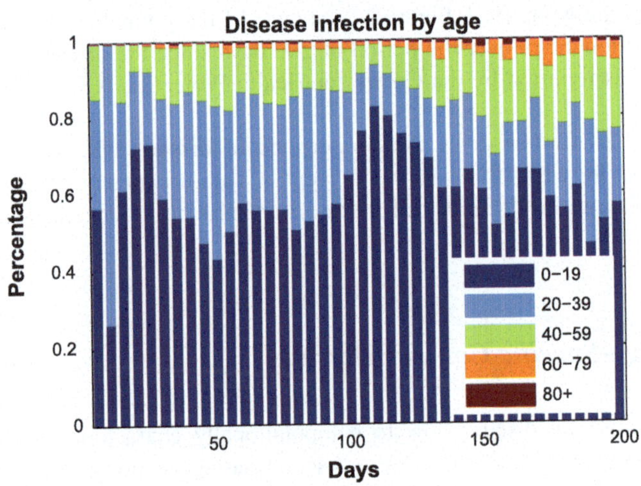

(b)

Fig. 2.2 The 2009 Hong Kong H1N1 influenza epidemic. The confirmed cases of the H1N1 infection reported by the Centre for Health Protection (CHP) of Hong Kong for the first 200 days since the disease onset. (**a**) The dynamics of disease transmission in terms of the daily number of newly infected cases. (**b**) The proportions of reported infections in different age subpopulations

represent children mixing with adults, mainly in households, and a wider contact "plateau" among adults, which accounts for contacts occurring in workplaces.

Individuals' cross-age contacts exhibit specific patterns that correspond to the likelihoods of individuals' mixing together within certain social settings (school, household, workplace, and general community), which in turn depend on the socio-demographic structure of the population (age distribution, school attendance, household size, and working population). These patterns suggest that disease transmissions through social contacts mainly occur in certain typical social contact settings. In addition, individuals' social contacts may change due to either their self-initiated behaviors (e.g., avoidance of public places) or governmental compulsory policies (e.g., school closures and workplace shutdowns).

Based on the above empirical findings, a reasonable strategy is to devise a computational means for directly inferring individuals' setting-specific contact patterns from the demographic structure of the considered population. Then, these setting-specific patterns can be used to estimate the overall social contacts that account for disease transmissions, by combining the patterns via their respective coefficients, to reveal the proportions of individuals' contacts within different social settings.

Specifically, we can define and compute the contact frequency between a pair of individuals in age groups i and j, i.e., c_{ij}, as the total number of contacts between two age groups, $C_{ij} = C_{ji}$, divided by the product of their population sizes, N_i and N_j:

$$c_{ij} = \frac{C_{ij}}{N_i N_j} = \frac{C_{ji}}{N_j N_i} = c_{ji} \tag{2.6}$$

Note that matrix $\mathbf{C}$ with the elements of c_{ij} is the overall contact matrix, which describes individuals' cross-age contact frequencies. Based on this definition, matrix $\mathbf{C}$ is symmetric for $c_{ij} = c_{ji}$.

Next, we calculate the probability for individuals of different ages mixing within certain social settings, i.e., individuals sharing the same places, namely households, schools, workplaces, and general communities. Then, we generate four matrices accounting for the specific patterns of individuals' contacts within each social setting, which are represented by $\mathbf{C}^H$ for contacts within households, $\mathbf{C}^S$ for schools, $\mathbf{C}^W$ for workplaces, and $\mathbf{C}^G$ for general communities. Thus, we can estimate the overall matrix of individuals' cross-age contact frequencies as a linear combination of the four setting-specific matrices:

$$\mathbf{C} = r^H \mathbf{C}^H + r^S \mathbf{C}^S + r^W \mathbf{C}^W + r^G \mathbf{C}^G \tag{2.7}$$

where the coefficients r^H, r^S, r^W, and r^G denote the proportions of individuals' effective contacts occurring in the above-mentioned social settings, respectively, and sum to one:

$$r^H + r^S + r^W + r^G = 1 \tag{2.8}$$

For illustration, we infer individuals' setting-specific contact patterns from the available census data of Hong Kong by calculating the likelihoods of individuals' mixing within different social settings [97]. The generated setting-specific contact pattern matrices are shown in Fig. 2.3.

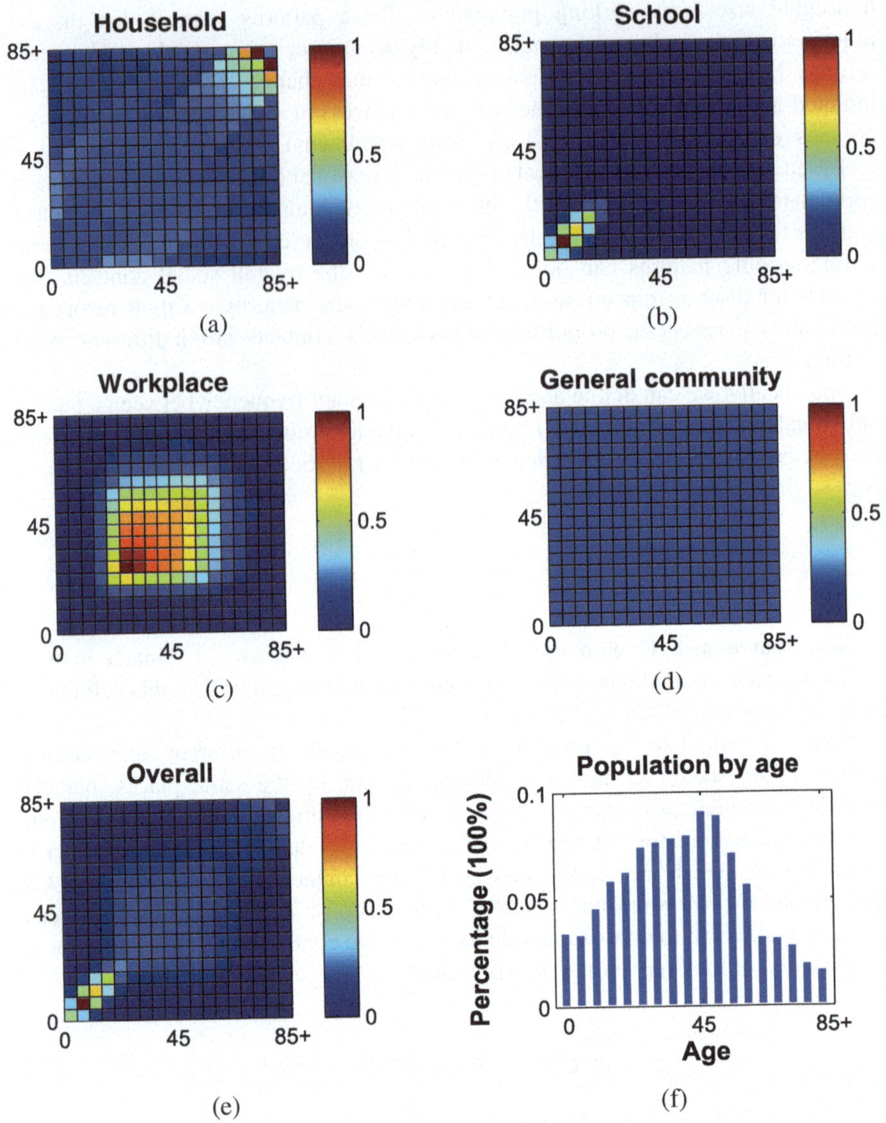

Fig. 2.3 Contact patterns inferred from the census data of Hong Kong. We consider disease transmissions among individuals between 0 and 85+ years old and divide these into 18 age groups. The contact matrices are generated corresponding to the likelihoods of individuals' mixing within respective social settings: household ($\mathbf{C}^H$), school ($\mathbf{C}^S$), workplace ($\mathbf{C}^W$), and general community ($\mathbf{C}^G$). The overall contact matrix is calculated as the linear combination of the four setting-specific contact matrices. The combination coefficient of each matrix denotes the proportion of effective contacts occurring in that social setting

Specifically, Fig. 2.3a describes the contacts in households, in which the main diagonal and two secondary diagonals correspond to the contacts within families and between parents and children. Figure 2.3b shows the pattern of contacts in schools, in which the strong diagonal elements among individuals below 20 years old indicate that students are more inclined to have contact with same-age individuals. Figure 2.3c presents the pattern of contacts in workplaces, in which the contacts are more frequent among individuals aged between 20 and 65. Figure 2.3d gives the pattern of individuals' random contacts with each other in general communities. We normalize the elements of the four generated contact pattern matrices, such that their total numbers of contacts are equal. For the overall contact matrix (see Fig. 2.3e), the coefficients used for combining the setting-specific matrices, r_{Φ} and $\Phi \in \{H, S, W, G\}$, can be approximately estimated as the fraction of disease infections occurring in the respective social settings.

Based on the population size of each age group (Fig. 2.3f), it has been shown that 31% of infections during the 2009 Hong Kong H1N1 influenza epidemic occurred in households [98]. In addition, we assume that the other three contact matrix coefficients are identical to those empirical estimations for the respective social settings [99, 100], as follows: 0.24 in schools, 0.16 in workplaces, and 0.29 in general communities.

2.3.3 *Validation*

To examine our aforementioned computational method, we now revisit the real-world scenario of the 2009 Hong Kong H1N1 influenza epidemic [101, 102]. Based on the census data of Hong Kong, we divide the host population between 0 and 85+ years of age into 18 groups to represent the individuals' age-specific contact patterns. In addition, we estimate the reproduction number to be $R_0 = 1.5$ in the initial stage of disease transmission [103]. The infectivity is $\alpha = 1.0$, which is homogeneous for all age groups. The susceptibility is estimated to be $\beta = 2.6$ for individuals below 20 years old, who are more susceptible than the rest of the population. The duration of H1N1 influenza infection is set to be 3.2 days [104]. Therefore, the recovery rate is calibrated as $\gamma = 0.312$ (i.e., 3.2^{-1}day^{-1}) based on an assumption that there is an exponential distribution of individuals' recovery from disease infection.

Based on the above parameterization, we now validate our epidemic model by comparing the model predictions with the real-world observations in terms of the daily new infection cases and the age-specific infection rates, as shown in Fig. 2.4. In doing so, we use the confirmed daily cases of H1N1 infection, as reported by the CHP of Hong Kong for the first 200 days since the disease onset in early May 2009, in which the infected cases were classified into five age groups.

The results of the simulation over the specified period show that disease infection peaked around 120 days after the disease onset, as shown in Fig. 2.4a. The young

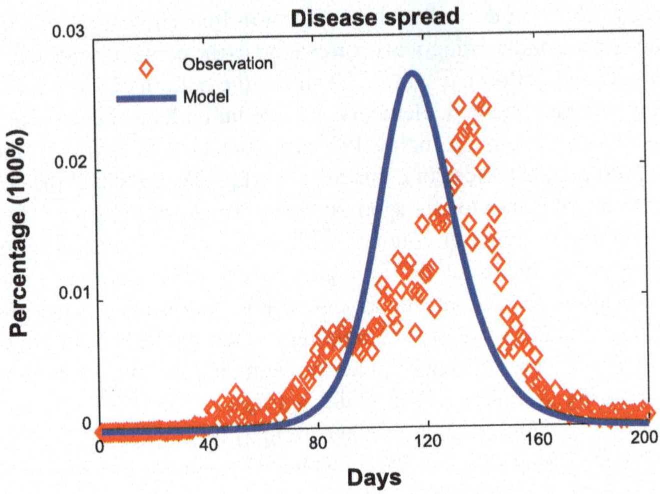

(a)

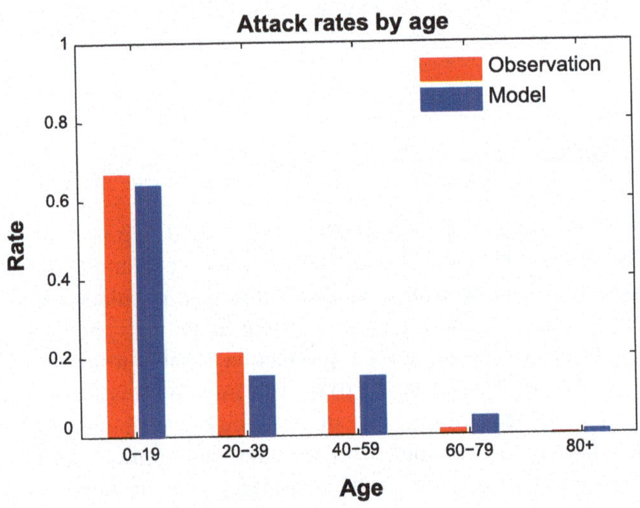

(b)

Fig. 2.4 The baseline dynamics of disease transmission. We calibrate our epidemic model based on the real-world scenario of the 2009 Hong Kong H1N1 influenza epidemic, as reported by the Centre for Health Protection (CHP) of Hong Kong for the first 200 days since the disease onset. (**a**) The dynamics of disease transmission in terms of the daily number of newly infected cases reported as a proportion of the total number of infections. (**b**) A comparison of the observation and model predictions in terms of disease infection rates by age and school-age students between 0 and 19 years old constituted a large proportion of the infection cases, while that of adults was relatively small, as shown in Fig. 2.4b.

2.4 Further Remarks

In infectious disease control, the effectiveness of intervention relies heavily on accurate predictions about where and when outbreaks may happen, and in what specific population groups. Such spatial, temporal, and demographic patterns of disease transmission are determined by the biological properties of a pathogen, i.e., its infectivity and transmissibility, as well as the demographic and social characteristics of a host population, i.e., individuals' vulnerability and their contact relationships. In view of this, public health authorities increasingly turn to mathematical models to characterize the dynamics of disease transmission and to evaluate the effect of adopted intervention strategies.

For example, compartmental models utilize deterministic differential equations to describe disease transmission in a group of randomly mixed individuals as a process of population transitions among different compartments, i.e., susceptible, infectious, and recovered. Social network models represent individuals' contact relationships using a network structure, and thereafter, model disease transmission as a stochastic percolation process. Based on epidemic models, strategies for controlling infectious diseases, e.g., antivirals [105], case isolation/quarantine [106], vaccination [107], and school and workplace closures [108], can be computationally simulated and the effectiveness of disease control further evaluated.

Accurate descriptions of individuals' contact relationships are essential for the utilization of epidemic models to characterize the dynamics of disease transmission, as such relationships affect the routes of disease transmission. In previous studies, individuals' contact relationships have been mainly characterized by utilizing statistical means to collect and analyze empirical data, e.g., individuals' contact activities in certain places/regions over a fixed time period [109]. For example, survey-based studies may provide static and empirical descriptions of the routes of disease transmission with respect to individuals' contact relationships. In this case, individuals' behavioral changes, i.e., potential responses to disease transmission and adopted interventions, are not taken into account, although they significantly affect the disease transmission during an epidemic [110].

However, in the real world, it is difficult to collect real-time data to characterize behavioral changes throughout the period of an epidemic. In fact, the lack of realistic data to characterize the disease transmission in a host population has presented a major challenge to effective disease intervention. To overcome this, we resort to developing computational methods for inferring contact patterns that incorporate individuals' social, demographic, and behavioral characteristics, and accordingly evaluate the effects of these patterns on the disease dynamics.

2.5 Summary

In this chapter, we introduced a number of key concepts and computational models that are essential for characterizing infectious disease dynamics, while adequately addressing the heterogeneity of a host population. In doing so, we addressed the host population heterogeneity in terms of age-specific susceptibility, infectivity, and the contact frequency among different population groups. Furthermore, we developed a specific compartmental model to describe the dynamics of influenza-like disease transmission in an age-structured host population.

In order to characterize individuals' contact relationships, we decomposed individuals' disease transmission-related contacts into those within four specific social settings: school, household, workplace, and general community. Accordingly, the coefficient for each contact matrix represents the proportion of individuals' contacts occurring in that social setting. In doing so, we utilized a computational method to infer individuals' setting-specific contact matrices from the socio-demographic data of the population.

Finally, to demonstrate our computational method, we conducted simulation-based experiments based on the real-world 2009 Hong Kong H1N1 influenza epidemic. The results showed that individuals' contacts, as inferred from the considered host population, exhibited clear age-specific patterns. Based on the analysis, we further validated our epidemic model by comparing the temporal-demographic patterns of disease transmission, i.e., the daily new infection cases and age-specific infection rates, between our model predictions and the real-world observations. This work provides the basis for our further discussion of vaccination programs in the following chapters.

Chapter 3
Strategizing Vaccine Allocation

In the control and prevention of infectious diseases, one of the most effective measures is vaccination. A key question for public health authorities is the optimal allocation of a finite number of available vaccine doses to most effectively reduce disease incidence. This is closely related to the question of which age groups or subpopulations will be most vulnerable and should be vaccinated first. Thus, if we can answer this question, we will be in a strong position to combat infectious diseases.

In this chapter we focus on the above questions. Building on the age-structured compartmental model of infectious diseases introduced in the preceding chapter, we discuss a computational means for identifying and prioritizing the target subpopulations for effective vaccine allocation [55]. Through a series of simulation-based experiments, we examine the performance of such a vaccine allocation strategy by considering different epidemic scenarios as well as other intervention strategies.

3.1 Vaccination

Vaccination has long been recognized as one of the most effective methods for the control and prevention of infectious diseases. In this section, we review the basic notions and existing methods of vaccination. Specifically, we discuss the principal idea behind vaccination, in terms of achieving the herd immunity effect. We also look at some of the existing vaccination strategies that have been designed for disease control and prevention.

J. Liu, S. Xia, *Computational Epidemiology*, Health Information Science,
https://doi.org/10.1007/978-3-030-52109-7_3

3.1.1 *Herd Immunity*

Generally, the effect of vaccination is judged in two respects: (1) how effectively it can directly immunize vaccinated individuals; and (2) how effectively it can indirectly protect unvaccinated individuals from getting infected by reducing disease transmission. To discuss effectiveness, we need to further introduce the concept of herd immunity, which is often mentioned in the literature although with several slightly different meanings. Here, we adopt the definition as given by John et al. [111], which is that herd immunity refers to "the proportion of subjects with immunity in a given population," which may be due to natural recovery from infection, vaccine-induced immunization, or a combination of both. In essence, herd immunity indicates the proportion of immunized individuals in a host population who will be able to resist the spread of a disease. Due to herd immunity, it will not be necessary for everyone in a host population to be vaccinated to prevent outbreaks, because the size of the susceptible population has been reduced, and so has the probability of disease transmission from the infectious to the susceptible. This indirect protection of unvaccinated individuals, as a result of vaccine-induced herd immunity, is often referred to as the herd immunity effect, which is schematically illustrated in Fig. 3.1.

Next, we introduce the notion of a herd immunity threshold. This indicates the critical portion of a host population that must be vaccinated/immunized to prevent disease transmission. Studies on a herd immunity threshold have focused primarily on how to determine the minimum vaccination coverage needed to prevent a potential outbreak. It should be noted that in epidemiology, the prevention of an outbreak entails that the average number of secondary infections per infectious individual should be less than one, which is equivalent to saying the basic reproduction number $R_0 < 1.0$ (see also Sect. 2.1.1). In this regard, the concept of a herd immunity threshold reflects the vaccination coverage that leads to $R_0 = 1.0$. In a standard SIR model, which assumes that individuals of a host population have uniform and random contacts with each other, the herd immunity threshold, denoted by θ, for random vaccination (assuming 100% vaccine effectiveness), can readily be written as follows:

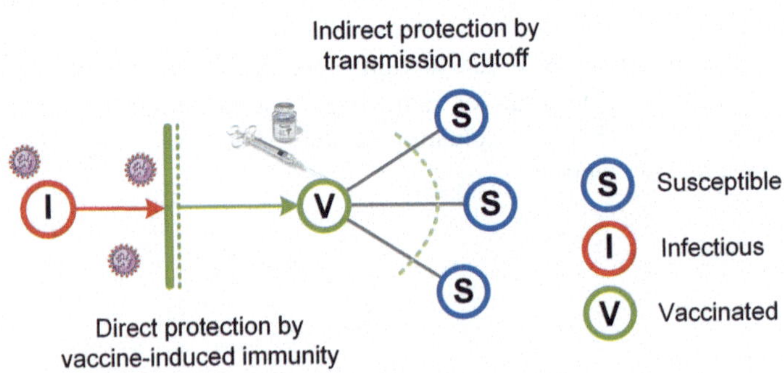

Fig. 3.1 The effect of vaccine-induced herd immunity

$$\theta = 1 - \frac{1}{R_0} \tag{3.1}$$

As can be seen, the usefulness of the notion of the herd immunity threshold is that it provides a way to examine the effectiveness of vaccination programs for disease control, that is, to evaluate the level of vaccination coverage in a host population.

3.1.2 Vaccine Allocation Strategy

As aforementioned, vaccination coverage in a host population plays a significant role in ensuring the effectiveness of a vaccination program. Due to limited vaccine supply, providing adequate and timely vaccine doses remains an open challenge, especially when encountering emerging infectious diseases. In this case, public health authorities in charge of vaccination programs need to address the problem of how to allocate a finite number of available vaccine doses for most effectively controlling disease transmission.

Public health authorities in different countries may adopt different vaccination prioritization plans in situations of limited vaccine supply. The WHO generally recommends that the priority for vaccine-based immunization should be given to essential service providers and individuals at a high risk of death and severe complications [37]. In addition to front-line workers and high-risk individuals, another important group of the susceptible population is healthy adults and children, which can be very large in size [112–114]. Effectively vaccinating ordinary susceptible individuals can help build up the population-level herd immunity and hence can directly contribute to decreasing disease transmission [115, 116]. In this regard, we focus on vaccine allocation strategies for specifically targeting the ordinary susceptible population to enhance population-level herd immunity.

Previous studies on modeling vaccine allocation have been focused on mathematical optimization approaches, and aimed at optimizing the predicted outcomes of disease control. In this context, the results of optimal vaccine allocation depend on the considered outcome measures and the projected time intervals. For example, some researchers have suggested that vaccine doses should be allocated to high-risk groups, thereby reducing influenza-attributed morbidity and mortality [117, 118]. Meanwhile, other researchers suggest that disease transmission could be more effectively reduced by allocating vaccine doses to the population groups with a higher possibility for disease transmission, e.g., school children, instead of the high-risk groups [119, 120].

Regarding vaccination during the different stages of disease transmission, Medlock et al. pointed out that optimal vaccine allocation should take into account disease dynamics and the vaccine availability [121]. Matrajt et al. observed that the optimal allocation of vaccines, e.g., with 25% vaccination coverage, might involve switching between allocating vaccine doses to transmitting groups to allocating vaccine doses to vulnerable groups at the approximate peak of an outbreak [122].

Similar results for vaccination in different stages were also reported by Myliusa et al. [123] and Bansal et al. [124].

Generally, the methods outlined in the preceding chapter adopt a fixed objective function to be optimized and an optimal strategy to be computed for the given period of time. To function, these methods make several assumptions. Specifically, they assume we are able to make accurate predictions about disease dynamics, and that we know in advance the number of available vaccine doses (needed to set up the constraints) and the timing of vaccine release (needed to determine the optimization time period). In real-life applications, such assumptions are evidently rather strong, as we generally have limited knowledge about disease-associated epidemiological features, e.g., infectivity and transmissibility, during the spread of an emerging disease. In addition, we may not know the quantity and timing of vaccine supply. As a result, the optimization-based methods may not be practically feasible.

Several recent studies have been performed to address the vaccine allocation problem by focusing on disease transmission rather than on future predictions. Importantly, the authors of these studies allowed vaccine allocation to be dynamically/adaptively adjusted in relation to the dynamics of disease transmission, e.g., based on the time-varying, age-specific incidences of infection revealed from surveillance data [125], or the real-time monitoring of hospitalization and infection-induced death [126]. The findings of these studies have partly solved the problem of obtaining detailed knowledge of an emerging disease (i.e., the parameters used in an epidemic model). For example, proxy indicators may be used for determining vaccine allocation, e.g., the group-specific confirmations of infection, or the hospitalization and mortality rates [127].

So far, in this section we have reviewed some of the key concepts and methods used in planning and evaluating the effectiveness of vaccination programs in terms of vaccination coverage in a host population. Another important issue is that the actual coverage of a vaccination program could be affected by individuals' willingness to be vaccinated (for more details, see Chaps. 4–6).

3.2 Vaccination Priorities

In a real-world situation, the vaccine supply is usually limited. This is partly because the quantity of available vaccine doses may be insufficient to meet the actual demand, and partly because the vaccine's release after the disease onset may be delayed. Thus, an important concern for public health authorities is to achieve the goals of vaccination by making the best use of the finite number of available vaccine doses. For example, in the United States, the National Vaccine Advisory Committee (NVAC) and the Advisory Committee on Immunization Policy (ACIP) have set the goals of vaccination as being to weaken health effects, including severe morbidity and death, and minimize socio-economic effects [128]. Accordingly, the NVAC and ACIP have recommended a priority vaccination for vaccine workers, health-care providers, and the ill elderly, followed by healthy people aged 2–64 [129].

In this book, we consider vaccine allocation for the purpose of most effectively minimizing disease transmission in the whole population. In this case, the key problem is how to determine the relative vaccine allocation priority for each of the subpopulations corresponding to their respective roles in disease transmission. That is, the challenge for effectively allocating a finite number of vaccine doses is herein translated into the question of how to adaptively adjust vaccine allocation to various subpopulations with reference to the immediate situation of disease transmission, such as the disease incidence and prevalence rates in different subpopulations, the possible changes of individuals' contact relationships, and the total number of available vaccine doses and the time of vaccine release.

3.3 Age-Specific Intervention Priorities

During the 2009 Hong Kong H1N1 influenza epidemic, the government announced the immediate closure of all primary schools and kindergartens when the first non-imported case was confirmed, and at the same time, targeted children between the ages of 6 months and 6 years as the priority groups for vaccination [130]. Relevant to this, a pertinent question is how we can systematically determine the relative priorities of subpopulations for disease interventions during the time course of disease transmission, where such intervention measures may include vaccine allocation, contact reduction, or a combination of both.

The aim of our work is to provide a method to enable optimal prioritization of subpopulations for disease interventions, using a combination of age-specific vaccine allocation with setting-specific contact reduction. We evaluate the effects of disease interventions for containing disease transmission by measuring the reduction in the reproduction number. By doing so, we show which subpopulations should be prioritized for vaccination, so as to generate the greatest reduction in the number of disease transmissions. We do so by considering the marginal effects of reducing the reproduction number for different cases of vaccine allocation by age, and reducing contacts by social setting.

For demonstration, we use a compartmental model to describe the dynamics of an influenza-like disease transmission in an age-structured host population. We parameterize the epidemic model with the epidemiological data from the 2009 Hong Kong H1N1 influenza epidemic, and further implement our method to identify the relative priorities of subpopulations for disease interventions in Hong Kong. Additionally, we carry out a series of simulation-based experiments with different settings of disease intervention.

Compared with existing optimization-based approaches, our method has the following features:

- The strategy utilizes prior knowledge about individuals' age-specific susceptibility and infectivity, real-time disease prevalence in each of the subpopulations, and the basic patterns of individuals' cross-age contact frequency within each social setting. Moreover, it does not rely on detailed information about individuals' actual contacts, nor their potential behavioral changes in response to disease transmission, which would be difficult to determine rapidly and accurately.

- The strategy is designed to most effectively reduce disease transmission by optimizing the allocation of a finite number of vaccine doses to certain target subpopulations. Crucially, these identified vaccination priorities can be adaptively regulated based on the dynamics of disease transmission. That is to say, the number of vaccine doses suggested to be allocated to each subpopulation can be dynamically adjusted according to the latest progress of disease transmission and vaccine supply.
- The developed vaccine allocation method also incorporates the effects of other disease intervention measures being implemented simultaneously with vaccination, e.g., individuals' contact reductions. Therefore, in a situation of integrated disease interventions, our method can provide more accurate and effective solutions for vaccine allocation.

3.3.1 *Modeling Prioritized Interventions*

We now introduce the standard epidemic model and our detailed computational method for identifying the relative priorities of age-specific subpopulations for intervention measures.

First, we use vector $\mathbf{I}(k) = [I_1(k), \ldots, I_N(k)]^T$ to denote the number of infectious individuals in each of N age groups at the kth generation of disease infection. Then, we characterize the dynamics of disease infection from generation k to generation $k+1$ as follows:

$$\mathbf{I}(k+1) = \mathbf{K}\mathbf{I}(k) \tag{3.2}$$

where $\mathbf{K}$ is the reproduction matrix, also known as *next generation matrix* (NGM) [131].

For the earlier-mentioned SIR model, the reproduction matrix $\mathbf{K}$ can be written as follows:

$$\mathbf{K} = (\mu\gamma^{-1})\mathbf{SBCA} \tag{3.3}$$

where matrix $\mathbf{S}$ describes the size of the susceptible population in each age group; it has elements $S_1, S_2, \ldots, S_N$ in the diagonal and zeros elsewhere. Matrix $\mathbf{B}$ summarizes individuals' age-specific susceptibility in the diagonal elements of $\beta_1, \beta_2, \ldots, \beta_N$, and zeros elsewhere. Matrix $\mathbf{A}$ gives the age-specific infectivity of infected individuals with the diagonal elements of $\alpha_1, \alpha_2, \ldots, \alpha_N$, and zeros elsewhere. Matrix $\mathbf{C}$, known as the contact matrix (see Eq. (2.7) in Sect. 2.3.2), describes the frequency of contacts between two age groups. Moreover, during the course of disease transmission, the susceptible populations will decrease in size over time, and therefore matrices **S** and **K** will change dynamically.

In epidemiology, the effective reproduction number R_t refers to the number of new infection cases caused by a typical infectious individual in a completely susceptible population. By constructing the NGM, the effective reproduction number R_t in the context of an age-structured host population can be approximately estimated as the dominant eigenvalue of reproduction matrix **K**:

$$R_t = \rho(\mathbf{K}) \tag{3.4}$$

Given the condition that matrices **S**, **B**, **C**, and **A** are all symmetric, R_t can be approximately calculated as:

$$R_t = \boldsymbol{x}_1^T \mathbf{K} \boldsymbol{y}_1 \tag{3.5}$$

where $\boldsymbol{x}_1$ and $\boldsymbol{y}_1$ are the top left and right eigenvectors of reproduction matrix **K** (the corresponding top eigenvalue is R_t). Specifically, we choose a normalized format of each eigenvector in which the elements are positive and sum to one. As proposed by Wallinga et al. [106], $\boldsymbol{y}_1$ and $\boldsymbol{x}_1$ approximately correlate to the number of new infections in each age group, **I**:

$$\begin{aligned} \boldsymbol{y}_1 &\propto \mathbf{I} \\ \boldsymbol{x}_1 &\propto \mathbf{S}^{-1}\mathbf{B}^{-1}\mathbf{A}\mathbf{I} \end{aligned} \tag{3.6}$$

Therefore, we can describe disease transmission (effective reproduction number R_t) with reference to the current disease prevalence, the susceptible population sizes, and their cross-age contact frequencies. In the following subsections, we examine the effects of disease interventions (vaccination and contact reduction) on containing disease transmission, by measuring these interventions' marginal reductions of R_t.

3.3.2 *Effects of Vaccination*

As aforementioned, disease transmission can be estimated based on R_t, which correlates to reproduction matrix **K**. Thus, the change of the effective reproduction number, dR_t, can be calculated from $d\mathbf{K}$ as follows:

$$dR_t = \boldsymbol{x}_1^T d\mathbf{K} \boldsymbol{y}_1 \tag{3.7}$$

As vaccination that immunizes susceptible individuals can reduce the sizes of susceptible populations in different age groups, the reduction of R_t due to vaccination will be proportional to the following term:

$$dR_t \propto \boldsymbol{x}_1^T(d\mathbf{S})\mathbf{BCA}\boldsymbol{y}_1 \tag{3.8}$$

Specifically, when targeting a susceptible population in age group i, the effects of vaccination can be calculated as follows:

$$\frac{dR_t}{dS_i} \propto \boldsymbol{x}_1^T\left(\frac{d\mathbf{S}}{dS_i}\right)\mathbf{BCA}\boldsymbol{y}_1 \tag{3.9}$$

By combining the elements of each matrix, we arrive at an indicator that evaluates the marginal reduction of R_t by vaccinating a unit of susceptible individuals in age group i:

$$\frac{dR_t}{dS_i} \propto \frac{\alpha_i}{\beta_i}\left(\frac{I_i}{S_i}\right)^2 \tag{3.10}$$

Due to the lack of knowledge about the reporting rate of disease infections, i.e., the ratio of confirmed cases to the overall infections, we approximate the susceptible population size S_i using the population size N_i, based on the assumption that the number of infections is relatively small in the host population. Therefore, we can prioritize each age group as follows:

$$\frac{dR_t}{dS_i} \propto \frac{\alpha_i}{\beta_i}\left(\frac{I_i}{N_i}\right)^2 \tag{3.11}$$

Therefore, with the above equation, we can determine the relative priorities of age groups for vaccine allocation with respect to their age-specific infectivity, susceptibility, population sizes, and current disease prevalence.

3.3.3 *Effects of Contact Reduction*

To contain disease transmission by reducing the effective contacts in a host population, we examine the change of the effective reproduction number, dR_t, with respect to the reduction of individuals' contact frequencies, $d\mathbf{C}$:

$$dR_t \propto \boldsymbol{x}_1^T\mathbf{SB}(d\mathbf{C})\mathbf{A}\boldsymbol{y}_1 \tag{3.12}$$

We focus on the effects of contact reduction in terms of reducing the number of individuals' contacts within a social setting, $\psi \in \{H, S, W, G\}$:

$$\frac{dR_t}{dr^{\psi}} \propto \boldsymbol{x}_1^T\mathbf{SBC}^{\psi}\mathbf{A}\boldsymbol{y}_1 \tag{3.13}$$

Then, based on the inferred setting-specific contact matrices, the relative priority for contact reduction that targets social setting ψ can be computed as follows:

$$\frac{dR_t}{dr^{\psi}} \propto \sum \left(\alpha_i I_i \sum \left(c_{ij}^{\psi} \alpha_j I_j\right)\right) \tag{3.14}$$

Thus, with the above equation, we will be able to further estimate the relative priorities of social settings for contact reduction, having taken into consideration individuals' age-specific infectivity, current disease prevalence, and setting-specific contact patterns.

3.3.4 Integrated Measures

Next, we consider the case of a campaign against the spread of an infectious disease, in which multiple intervention measures will be implemented at the same time. In view of this, we are interested in evaluating the effect of vaccination and contact reduction being implemented simultaneously. We estimate the marginal reduction of the effective reproduction number, $d^2 R_t$, corresponding to simultaneous vaccination of the susceptible population $d\mathbf{S}$ and the reduction of effective social contacts $d\mathbf{C}$:

$$d^2 R_t \propto \boldsymbol{x}_1^T (d\mathbf{S})\mathbf{B}(d\mathbf{C})\mathbf{A}\boldsymbol{y}_1 \tag{3.15}$$

Specifically, when selecting age group i for vaccination and social setting ψ for contact reduction, the effect of implementing the two intervention measures on reducing the effective reproduction number R_t can be evaluated as follows:

$$\frac{d^2 R_t}{dS_i dr^{\psi}} \propto \boldsymbol{x}_1^T \left(\frac{d\mathbf{S}}{dS_i}\right) \mathbf{B}\mathbf{C}^{\psi}\mathbf{A}\boldsymbol{y}_1 \tag{3.16}$$

That is,

$$\frac{d^2 R_t}{dS_i dr^{\psi}} \propto \frac{\alpha_i}{S_i} \sum \left(c_{ij}^{\psi} \alpha_j I_j\right) \tag{3.17}$$

Hence, by evaluating the interplay of the two intervention measures, we can identify the relative priorities of age groups and social settings when vaccine allocation and contact reduction are implemented simultaneously. That is, the number of vaccine doses allocated to each age group will be proportional to the relative priority of that group, and the contact reduction will target the social setting with the highest priority.

3.4 Case Study

In this section, we demonstrate our vaccine allocation strategy using the real-world scenario of the 2009 Hong Kong H1N1 influenza epidemic, by determining the relative priorities of the age-specific subpopulations for vaccination.

3.4.1 Hong Kong HSI Vaccination Programme

Figure 3.2 shows the H1N1 scenario in terms of the daily numbers of reported infections in different age groups during the spread of the disease in Hong Kong. Accordingly, Fig. 3.3 presents the relative priorities for vaccine allocation in those age groups during the course of disease transmission.

Generally, individuals between 0 and 29 years old are the most important subpopulation for containing disease transmission by vaccination. However, for each specific age group, the identified priorities vary between different stages of disease transmission. For the first month since the disease onset, up to day 25, we can observe that individuals between 0 and 29 years of age are targeted as the top priority subpopulation for vaccination. This describes the situation in which outbreaks will appear among school-age students due to their high frequency of contacts. Subsequently, on day 50, it can be observed that the relative priorities of individuals aged between 0 and 9 and those between 20 and 29 are increased, while the priorities of individuals between 10 and 19 are reduced. When disease

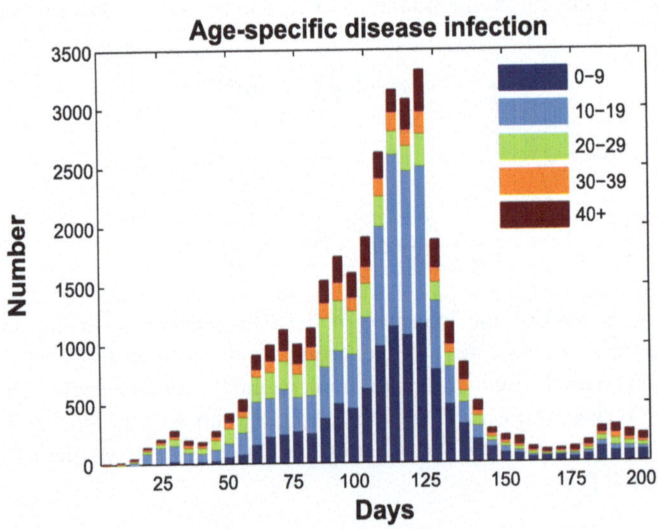

Fig. 3.2 The numbers of reported infections in different age groups during the 2009 Hong Kong H1N1 influenza epidemic

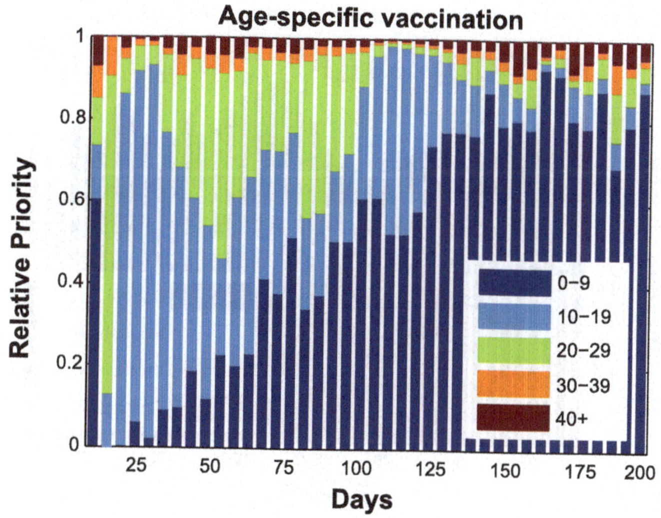

Fig. 3.3 Relative priorities of age-specific subpopulations for vaccine allocation during the course of disease transmission

infection peaks near day 120, individuals between 0 and 19 are predicted to become the dominant priority for vaccination, which agrees with the real-world observation that children and school-age students accounted for a large proportion of the new infections in this stage (Fig. 3.2). Finally, in the decay stage of the epidemic, children between 0 and 9 will become the subpopulation with the highest priority for vaccination. However, it should be pointed out that vaccination is more effective in the initial stage of disease transmission than in the stage of decay.

Figure 3.4 shows the relative priorities of the social settings for disease intervention by contact reduction among individuals. Overall, the reduction of individuals' contacts within schools is identified as the key measure for containing disease transmission throughout the whole period of disease transmission. Disease transmissions within households and workplaces account for a relatively large proportion during the initial and the decay stages of disease spread (between day 25 and day 100 and between day 150 and day 200, respectively), however, when infection peaks around day 120, interactions within these environments account for a relatively small proportion of disease transmissions. As mentioned before, the estimated proportion of infections occurring in schools was not the largest among the four considered social settings (i.e., the empirical estimation for schools was approximately 24%, and for households 31%). However, the disease infections that had already occurred in schools played a significant role in the disease transmission among the host population. Therefore, the reduction of individuals' effective contacts in the social setting of schools (through school closures as well as school sanitation and disinfection) should be implemented immediately after the disease onset, and conducted for the whole period of disease transmission.

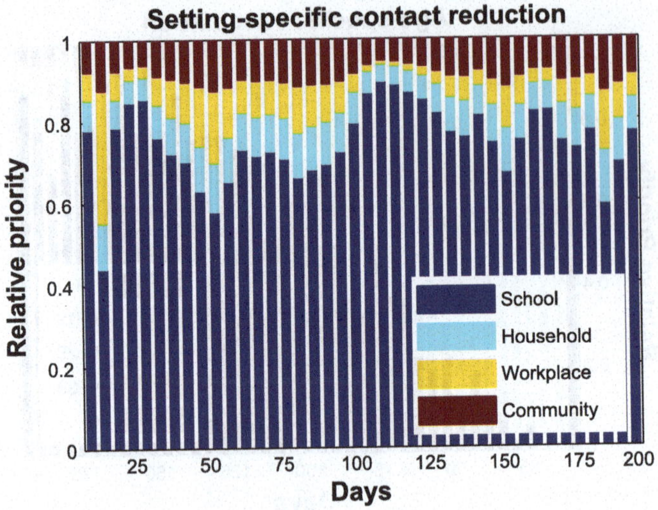

Fig. 3.4 Relative priorities of social settings for contact reduction during the course of disease transmission

Figure 3.5 further presents the relative priorities of the age groups and social settings when vaccine allocation and contact reduction are implemented simultaneously. Generally, vaccination of individuals between 0 and 19 and contact reduction in schools are the most important measures for containing disease transmission. For people of older ages, contact reduction in households, workplaces, and general communities is more important than contact reduction in schools. Specifically, as shown in Fig. 3.5a, on day 1 of disease transmission, contact reduction in schools and vaccination of individuals aged between 5 and 19 should be the top priority to contain disease transmission. On day 60, as indicated in Fig. 3.5b, individuals between 15 and 19 are identified as the target subpopulation for vaccination, followed by individuals aged 10–14 and 5–9. At this stage, contact reduction in the social setting of schools remains the top priority. When disease infection peaks at approximately day 120, as illustrated in Fig. 3.5c, the age groups 5–9 and 10–14 are predicted to become the most important subpopulations for vaccination. In the final stage of disease transmission on day 180, as shown in Fig. 3.5d, vaccinating children becomes more important.

By comparing the results of disease intervention using only vaccination (Fig. 3.3) with those using simultaneous vaccination and contact reduction (Fig. 3.5), we can observe that the age group 20–29 has a higher priority for vaccine allocation in the case of vaccination only than in the case of adopting both intervention measures. This is mainly due to the interplay of the two intervention measures, in that the effects of reducing effective contacts in schools can prevent or delay disease transmission to other age groups.

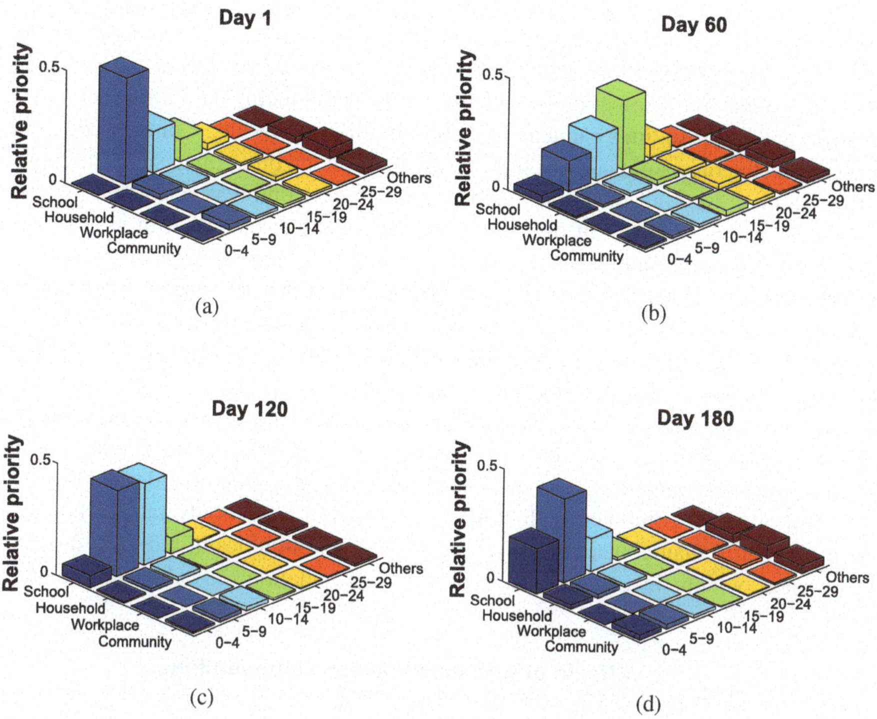

Fig. 3.5 Relative priorities of subpopulations and social settings for simultaneous implementation of age-specific vaccination and setting-specific contact reduction during the course of disease transmission

3.4.2 Effects of Prioritized Interventions

Next, to demonstrate the performance of our method, we carry out several simulation-based experiments to examine infectious disease transmission under different settings of vaccination and contact reduction. In the Hong Kong H1N1 influenza epidemic, the government closed all primary schools, kindergartens, and special schools immediately after the first local case was confirmed on June 10, 2009. Also, the Human Swine Influenza Vaccination Programme (HSIVP) was launched on December 21, 2009. As of early February 2010 (50 days after the vaccine became available), approximately 180,000 individuals had been administered vaccine doses, which accounted for approximately 2.5% of the overall population. For demonstration, we examine disease transmission under various intervention measures, i.e., *vaccination with a coverage of 2.5% of the host population* and *contact reduction within different social settings*. The intervention measures are set to be implemented on day 75.

The results in Fig. 3.6 show that vaccination and contact reduction can ameliorate disease prevalence by reducing the incidence rate at the peak of an outbreak. In the case of disease intervention by contact reduction only, we can clearly observe from the time course of the infectious population sizes that reducing individuals' contacts in different social settings can lead to distinctly different results. Contact reduction in schools (the blue solid curve) outperforms those in the other three social settings, in terms of preventing the occurrence of an infection outbreak and lowering the incidence rate at the peak of disease prevalence. Contact reduction in households (the red solid curve) and workplaces (the yellow solid curve) each has a similar effect on disease control and a better performance than that in general communities (the green solid curve). These results agree well with our previous prioritization of social settings for contact reduction, in which schools are identified as the top priority, followed by households and workplaces.

For the implementation of vaccination and contact reduction simultaneously, the results are shown as the dashed curves in Fig. 3.6. With contact reduction in schools, disease transmission would be almost eliminated (the blue dashed curve). Vaccination when combined with contact reduction in households and workplaces performs better than contact reduction in general communities in reducing disease prevalence, which, in turn, is still better than vaccination only. In addition, the

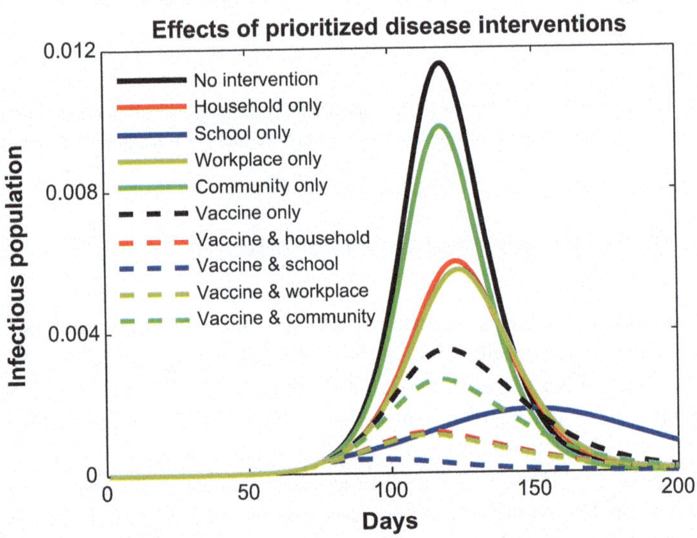

Fig. 3.6 Disease dynamics under the disease interventions of age-specific vaccination and setting-specific contact reduction. The baseline dynamics of disease transmission without any intervention (black solid curve); contact reduction only in schools (blue solid curve), households (red solid curve), workplaces (yellow solid curve), and general communities (green solid curve); vaccination only (black dashed curve); vaccination and contact reduction in schools (blue dashed curve), households (red dashed curve), workplaces (yellow dashed curve), and general communities (green dashed curve) (Color figure online)

implementation of contact reduction in schools only (the blue solid curve) leads to a lower incidence rate at the peak of disease outbreak than the simultaneous implementation of vaccination and contact reduction in general communities (the green solid curve), and also delays the time of disease outbreak.

3.5 Further Remarks

The two disease intervention measures studied here are age-specific vaccine allocation and social setting-specific contact reduction. Unlike previous studies in which statistical means are utilized to describe individuals' contact relationships, in this chapter we have discussed how to infer setting-specific contact matrices by decomposing individuals' actual disease transmission-related contacts into their contacts within four specific social settings, i.e., school, household, workplace, and the general community. Therefore, the contact-reduction method of disease intervention can be interpreted as the reduction of the proportion of individuals' contacts in a certain social setting. Furthermore, the changes of individuals' overall contacts can be interpreted as the changes of contact proportions within different social settings. We have also examined the reproduction number to evaluate the effects of implementing different intervention measures on containing disease transmission. Therefore, we can identify the priority subpopulations (in terms of age groups and social settings) based on the marginal reduction of the reproduction number caused by changes of the susceptible population sizes by age (vaccination) and of the contact proportions by social setting (contact reduction).

We have evaluated our method with respect to the real-world scenario of the 2009 Hong Kong H1N1 influenza epidemic and, thereafter, examined the relative priorities of subpopulations for age-specific vaccination and setting-specific contact reduction in Hong Kong. Our study has practical value for public health authorities in preparing and assessing their intervention measures for controlling an infectious disease. First, the age distribution of new infections will always be available from an epidemic surveillance system, e.g., the CHP in Hong Kong. Next, the basic patterns of individuals' contacts in different social settings will depend mainly on the socio-demographic characteristics of the population, which can be derived either through statistical means or by computational methods from the census data for the host population. Finally, disease control will be more effective when multiple intervention measures are implemented simultaneously.

It should be pointed out that the results from our method partly depend on the accuracy of the age distribution of new infections reported by the surveillance system. In addition, other potential factors that also affect the results may include the reporting rates of infection, which may vary for individuals in different age groups due to their physical and biological conditions, and the time needed for the case confirmations, which may lead to a delayed response to disease transmission.

3.6 Summary

In this chapter, we developed a solution to the problem of how to allocate a finite number of vaccine doses to an age-structured host population. To achieve this, we considered two measures: vaccination, which immunizes susceptible individuals in different age subpopulations, and contact reduction, which reduces individuals' effective contacts in different social settings. We presented a prioritization method to identify the target subpopulations that will lead to the greatest reduction in the number of disease transmissions when subjected to age-specific vaccination, or to vaccination integrated with contact reduction. We computed the relative priorities of subpopulations by considering the marginal effects of reducing the reproduction number in the cases of vaccine allocation by age and contact reduction by social setting. We demonstrated the performance of our method, based on the real-world 2009 Hong Kong H1N1 influenza epidemic, through a series of simulation-based experiments on disease transmission under different interventional settings.

Chapter 4
Explaining Individuals' Vaccination Decisions

Although vaccination has long been regarded as one of the most effective methods in controlling infectious diseases, public concerns about the safety and efficacy of vaccines will significantly affect the effective coverage of a vaccination program. In the real world, it remains highly challenging to achieve adequate and lasting vaccination coverage in a host population, especially for a voluntary vaccination program. An example is the decline of MMR vaccine uptake in Britain [132], subsequent to a controversial study associating the vaccine with the development of autism. Even though other studies contested this association, the public remained doubtful about the safety of the MMR vaccine, leading to a decrease in vaccine uptake, which was in turn followed by outbreaks of measles [132]. Similarly, public losses of confidence in the safety of pertussis vaccine have led to declines in its uptake in many countries, subsequently resulting in a series of large pertussis outbreaks [133].

In the remaining three chapters, we aim to explain how individuals make their decisions once a voluntary vaccination program becomes available, and what could be the dominant factors in their decision-making [53, 54]. Specifically, we focus our discussions on the following three key aspects:

- **Risks and Benefits of Vaccination**
  Vaccination immunizes susceptible individuals to prevent infection. However, vaccinated individuals also face the potential risk of adverse effects of the vaccine, which may sometimes lead to severe complications. In this regard, individuals making vaccination decisions typically focus on several determinants associated with the risks and benefits of vaccination, such as the perceived risk of disease infection, the perceived safety and efficacy of the vaccine, and the socio-economic costs associated with vaccination and disease infection (e.g., the monetary cost of vaccination, medical expenses for treatment in the case of infection, and possible absence from work[134]).

J. Liu, S. Xia, *Computational Epidemiology*, Health Information Science,
https://doi.org/10.1007/978-3-030-52109-7_4

- **Impact of Social Influences**
  In a socially connected population, an individual's behaviors or opinions are also subjected to the influence of his/her connected neighbors, i.e., social influences [135, 136]. That is to say, in a voluntary vaccination program, an individual's decisions are affected by the decisions of others. The social influences on individuals' vaccination decisions can derive from multiple sources, such as recommendations of friends or family members [137], suggestions from health professionals [138], and advice given by trusted colleagues [139].
- **Individuals' Subjective Perception**
  Individuals inevitably lack prior knowledge when balancing the risks and benefits of vaccination with a newly developed vaccine against an emerging infectious disease. In this case, vaccination decision-making relies on individuals' subjective perception of disease severity and vaccine safety. Specifically, individuals who are aware of the dangers of severe infectious diseases will tend to seek protection through vaccination. However, those who are conscious of the potential adverse effects of vaccination will be less willing to be vaccinated.

In this regard, there are two challenges for understanding the effectiveness of a voluntary vaccination program, as follows: (1) how to characterize individuals' vaccination decision-making, taking into consideration the above-mentioned factors; and (2) how to estimate the final population vaccination coverage that results from individuals' decision-making.

In what follows, we develop computational models to characterize individual-level vaccination decision-making corresponding to (1) the risks and benefits of vaccination; (2) the effect of social influences; and (3) individuals' subjective perception. We further examine the effect of individuals' vaccination decisions on ensuring adequate population vaccination coverage for effective infectious-disease control.

4.1 Costs and Benefits for Decision-Making

The perceived risks affecting individuals' decision-making are twofold: the risk of being infected for unvaccinated individuals, and the risk of a vaccine's adverse effects for vaccinated individuals. It has been found that individuals who worry about being infected or believe themselves vulnerable to infection will be more inclined to accept vaccination, and vice versa [140, 141]. Similar patterns have also been found for individuals' intention to be vaccinated when they felt that a pandemic would be severe and long-lasting [142]. Moreover, studies have shown that individuals with positive attitudes to vaccination, and who believe that vaccination can reduce their risk of infection, were motivated to be vaccinated [143, 144]. In contrast, individuals who worried about the vaccine's potential adverse effects and doubted its efficacy were found to have a lower level of vaccine uptake [145, 146].

Several mathematical models have been proposed to model individuals' vaccination decision-making, with these models utilizing payoff-based methods to characterize decision-making based on individuals' perceptions of the risks and benefits of vaccination. Bauch et al. characterized individuals' vaccination decisions as a modified minority game by exploring the herd immunity effect [147, 148]. In a related manner, game theory has been used to describe individuals' decision-making in favor of optimizing personal payoffs [149, 150]. Cojocaru extended the game-theoretic model by considering a finite number of heterogeneous population groups [151]. Perisic et al. further incorporated individuals' contact networks into the vaccination game-theoretic analysis [152, 153]. Moreover, some studies have considered social and psychological aspects of decision-making (e.g., social learning processes [154] and imitation behaviors [155–157]). Elsewhere, others have considered the problem of incomplete information, by addressing either the potential discrepancy between individuals' perception and real situations (e.g., perceived disease prevalence and adverse effects of vaccination [158]) or different sources of information [159, 160].

4.2 Game-Theoretic Modeling of Vaccination Decision-Making

Game theory provides a useful tool for characterizing the dilemma associated with voluntary vaccination. That is, for each individual, it would be better not to take the risk of vaccination, while benefiting from the herd immunity generated from the rest of the population keeping the vaccination coverage high. Game-theoretic analysis assumes that individuals possess perfect rationality to maximize their own gains by adjusting their decision of whether to vaccinate. If applied to a vaccination program, game theory implies that the host population's individuals will adjust their willingness to vaccinate by balancing the costs and benefits associated with the status of a spreading epidemic and the safety and efficacy of vaccines.

There are two types of costs associated with an individual's vaccination decision: (1) the cost of vaccination (e.g., the potential risk of a vaccine's adverse effects or the expense of vaccine administration); and (2) the cost of infection if not vaccinated (e.g., disease complications, expenses for treatment, or absence from work). We let ξ and ζ denote the costs associated with vaccination and infection, respectively, and use $\hat{\lambda}_i$ to represent the perceived risk of being infected by a susceptible individual i (see the definition of λ_i in Sect. 2.1.2). Then, we can introduce a cost function for individual i with a decision σ_i, as follows:

$$F_i(\sigma_i) = (1 + \sigma_i) \cdot \xi + (1 - \sigma_i) \cdot \hat{\lambda}_i \cdot \zeta \tag{4.1}$$

where ξ denotes the cost associated with being vaccinated, and $\hat{\lambda}_i \cdot \zeta$ denotes the cost associated with rejecting vaccination.

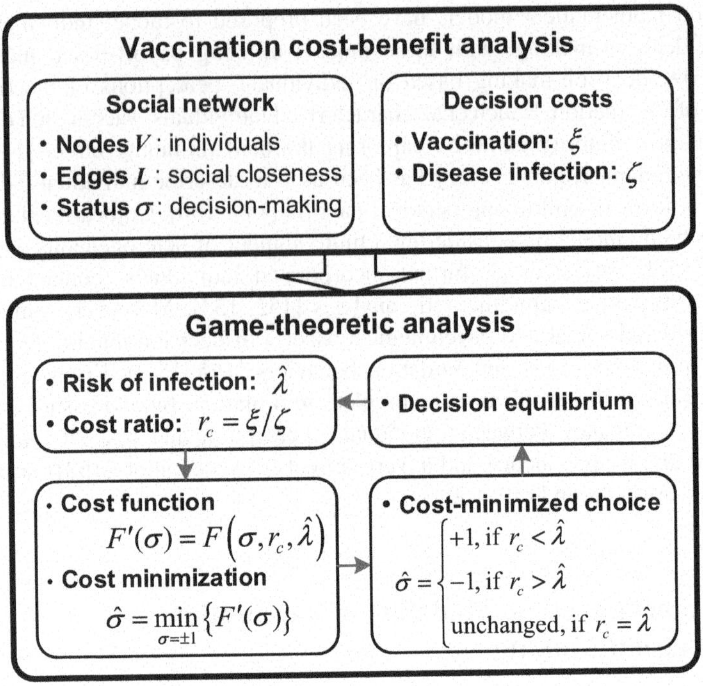

Fig. 4.1 A decision process for characterizing individuals' voluntary vaccination. A group of socially interconnected individuals can make decisions by minimizing all of the associated costs

Next, without loss of generality, we let $r_c = \xi/\zeta$ describe the ratio of ξ and ζ. Thus, we can further transform the cost function $F_i(\cdot)$ into the following:

$$F_i'(\sigma_i) = (1 + \sigma_i) \cdot r_c + (1 - \sigma_i) \cdot \hat{\lambda}_i \tag{4.2}$$

As illustrated in Fig. 4.1, we assume that individuals estimate the risk of disease infection based on their perception of disease severity, as reflected in the perceived disease transmission rate, $\hat{\beta}$, as well as their neighbors' vaccination decisions, as represented by N_i^{vac} and N_i^{non} for the numbers of neighbors who make the decision to vaccinate or not, respectively. In addition, vaccinated individuals are assumed to be successfully immunized from disease infection, while unvaccinated individuals can be infected and thus transmit the disease. Therefore, the perceived infection risk, $\hat{\lambda}_i$, can be computed corresponding to the proportion of unvaccinated neighbors as follows:

$$\hat{\lambda}_i = \hat{\beta} \cdot \left(\frac{N_i^{non}}{N_i^{vac} + N_i^{non}} \right) \tag{4.3}$$

Based on the above formulation, an individual can arrive at an optimal choice by minimizing the cost function in Eq. (4.2). In our model, individual i will accept vaccination ($\sigma_i = 1$) if $r_c < \hat{\lambda}_i$, reject vaccination ($\sigma_i = -1$) if $r_c > \hat{\lambda}_i$, and keep his/her decision unchanged from the previous step if $r_c = \hat{\lambda}_i$. We can express this cost-minimized choice of individual i, $\hat{\sigma}_i$, in the following form:

$$\hat{\sigma}_i = \begin{cases} +1, & \text{if } r_c < \hat{\lambda}_i \\ -1, & \text{if } r_c > \hat{\lambda}_i \\ \text{unchanged}, & \text{if } r_c = \hat{\lambda}_i \end{cases} \tag{4.4}$$

If all individuals follow the same strategy of minimizing their cost functions, after a certain number of iterations of decision-making a steady state will be reached, in which all individuals will have no incentive to change their decision in the next step.

4.3 Case Study

In the preceding section, we have shown how individuals' vaccination decision-making can be computationally modeled as an integrated decision process that incorporates their cost minimization with the effect of social influences. Next, we examine the individual-level decision-making process with respect to voluntary vaccination, and examine its potential effect on disease control using a real-world influenza epidemic as an example.

4.3.1 Hong Kong HSI Vaccination Programme

First, for our simulations, we calibrate the parameters of individuals' vaccination decision-making based on the 2009 H1N1 influenza epidemic (see also Sect. 2.3.1). To focus on the effect of social influences, we assume that the perceived disease transmission rate is equal to that of the actual disease transmission, i.e., $\hat{\beta} = \beta$. In addition, we construct a social network based on data of individuals' close proximity interactions (i.e., at distances less than 3 m) at a high school [69], where the social closeness w_{ij} between individuals i and j corresponds to the frequency of their interactions (the sum of all interactions between the two individuals during the day). The total number of nodes is $N = 788$ and the average node degree (the number of connected neighbors) is 35. The average link weight (social closeness) is 115 units. Based on our model parameterization, we carry out Monte Carlo simulations to experimentally study individuals' vaccination decision-making and the effect of the resulting vaccination coverage on disease control.

4.3.2 *Vaccination Coverage*

To investigate the impact of individuals' decision-making on the coverage, we first consider each individual's vaccination decision-making, with respect to three cases: (1) $R_0 = 1.2$; (2) $R_0 = 1.6$; and (3) $R_0 = 2.0$. Figure 4.2 shows the vaccination thresholds (see also Sect. 3.1.1) for eliminating the epidemic with respect to different basic reproduction numbers R_0.

Based on our decision model, we perform several simulations of vaccination dynamics to estimate the vaccination coverage at the steady state of individuals' decision-making. As shown in Fig. 4.3, we investigate the effect of individuals' decision-making on vaccination coverage, in which each individual tries to minimize his/her perceived cost. The level of vaccine uptake is determined by the cost ratio r_c and the perceived severity of disease infection, i.e., reproduction number R_0. Specifically, the simulation results show that the cost of vaccination (cost ratio r_c) fundamentally determines the resulting vaccination coverage, in that increasing the cost of vaccination will lower individuals' vaccination willingness (at the steady state of individuals' decision-making). In our considered situation, the vaccination coverage is greater than 70% when cost ratio $r_c = 0.1$, which means that the vaccination coverage is above the threshold for eliminating disease transmission. Gradually, as r_c increases and approaches 1.0, the vaccination coverage falls to 20%, which means voluntary vaccination may not be adequate to prevent disease outbreaks. In our model, when an individual perceives a sufficiently high severity of disease transmission (i.e., R_0 is increased from 1.2 to 2.0), the individual will

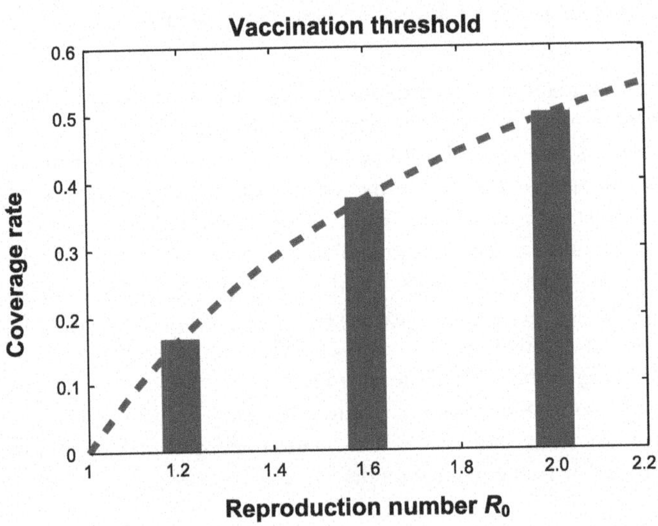

Fig. 4.2 The vaccination thresholds for preventing disease outbreaks, with respect to: (1) $R_0 = 1.2$; (2) $R_0 = 1.6$; and (3) $R_0 = 2.0$

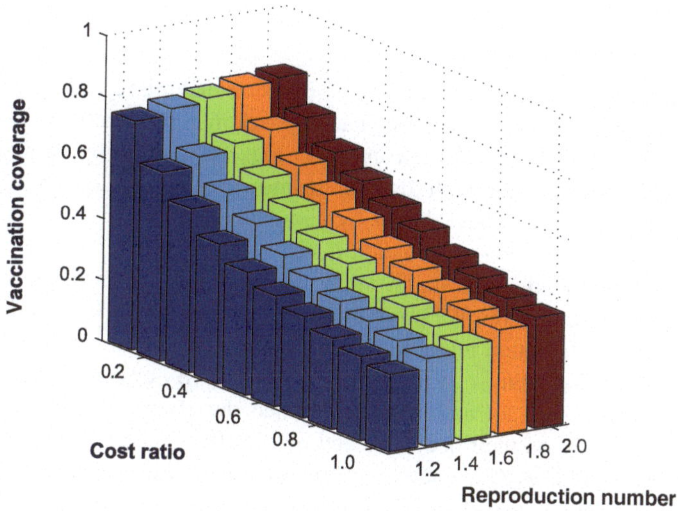

Fig. 4.3 Vaccination coverage at the steady state of individuals' decision-making. We investigate the effect of individuals' decision-making being motivated by cost minimization, with this effect measured by the resulting vaccination coverage, through varying the values of cost ratio r_c between 0 and 1.0. The disease severity (basic reproduction number R_0) is set as: (1) $R_0 = 1.2$; (2) $R_0 = 1.4$; (3) $R_0 = 1.6$; (4) $R_0 = 1.8$; and (5) $R_0 = 2.0$

change his/her previous choice of non-vaccination even if the cost of vaccination is high, due to his/her consideration of the risk of disease transmission.

4.4 Further Remarks

Game-theoretic analysis has been used in various studies to investigate the collective effect of individual-level vaccination choices on the population-level vaccination coverage, that is, the coverage of vaccination achieved at the equilibrium state of individuals' strategy adjustment. It has been found that the vaccination coverage resulting from individuals' collective self-interested choices will always be below the optimal level of vaccination for the whole population (i.e., the threshold of vaccination for the eradication and elimination of disease infection). Specific examples are the analysis of vaccination campaigns against smallpox [47], seasonal influenza [48], and more recently the H1N1 epidemic [161]. In the real world, individuals' choices deviate greatly from those that would result from perfect rationality.

4.5 Summary

In this chapter, we investigated individuals' decision-making regarding whether or not to be vaccinated in a voluntary vaccination program. Specifically, we considered individuals as rational decision-makers who make decisions by evaluating the risks and benefits associated with disease infection and vaccination. In this regard, we presented a decision model driven by game-theoretic analysis, in which each individual arrived at his/her vaccination decisions by minimizing their costs. By doing so, we provided an in-depth study of vaccination decision-making that integrates cost minimization. We used individuals' social networks to represent the structure of their interaction relationships and carried out a series of simulations of voluntary vaccination against an influenza-like disease. By evaluating the resulting coverage of voluntary vaccination, we examined the effect of individuals' cost-minimization motive on their decisions and on the effectiveness of disease control for the entire population (vaccination coverage), where individuals' choices were driven by the relative cost of the decision to vaccinate, i.e., cost ratio r_c. Based on the results in this chapter, we conclude that individuals' self-interested cost minimization will negatively affect the control of an infectious disease, by reducing the vaccination coverage. In this respect, our work can provide a means for estimating the effectiveness of a voluntary vaccination program.

Chapter 5
Characterizing Socially Influenced Vaccination Decisions

In addition to the self-estimated costs and benefits of vaccination against disease, as discussed in the previous chapter, individuals' vaccination decision-making is also subject to social influences, i.e., the vaccination decisions of their family members, friends, or colleagues. To better understand individuals' vaccination decision-making, in this chapter we further examine how people embark on an integrated decision-making process that incorporates both the cost-benefit analysis of vaccination decisions and the effect of social influences [53].

5.1 Social Influences on Vaccination Decision-Making

As illustrated in Fig. 5.1, we consider a group of individuals that make their vaccination decisions by both minimizing the associated costs and evaluating the decisions of others (social influences). Specifically, we focus on the social settings of individuals, which are structured with reference to their interaction relationships (interconnected individuals and their social closeness). Therefore, the effect of social influences among them will be heterogeneous with respect to the structure of their interactions. In addition, when individuals interact with those making similar choices, their decisions may be further affirmed; in the opposite case, their confidence in their decisions may be weakened [162]. In such a case, Social Impact Theory (SIT) provides a computational approach to characterize the effect of social influences corresponding to individuals' interactive relationships [163].

In outline, SIT describes how individuals change their attitudes or decisions in a structured social environment, and further suggests that the strength of the social effect is determined by the characteristics of the source (e.g., various attitudes or decisions), the closeness of their social relationships, and the number of sources holding similar attitudes or decisions. In this work, we provide an in-depth modeling framework for describing individuals' vaccination decision-making by integrating

J. Liu, S. Xia, *Computational Epidemiology*, Health Information Science,
https://doi.org/10.1007/978-3-030-52109-7_5

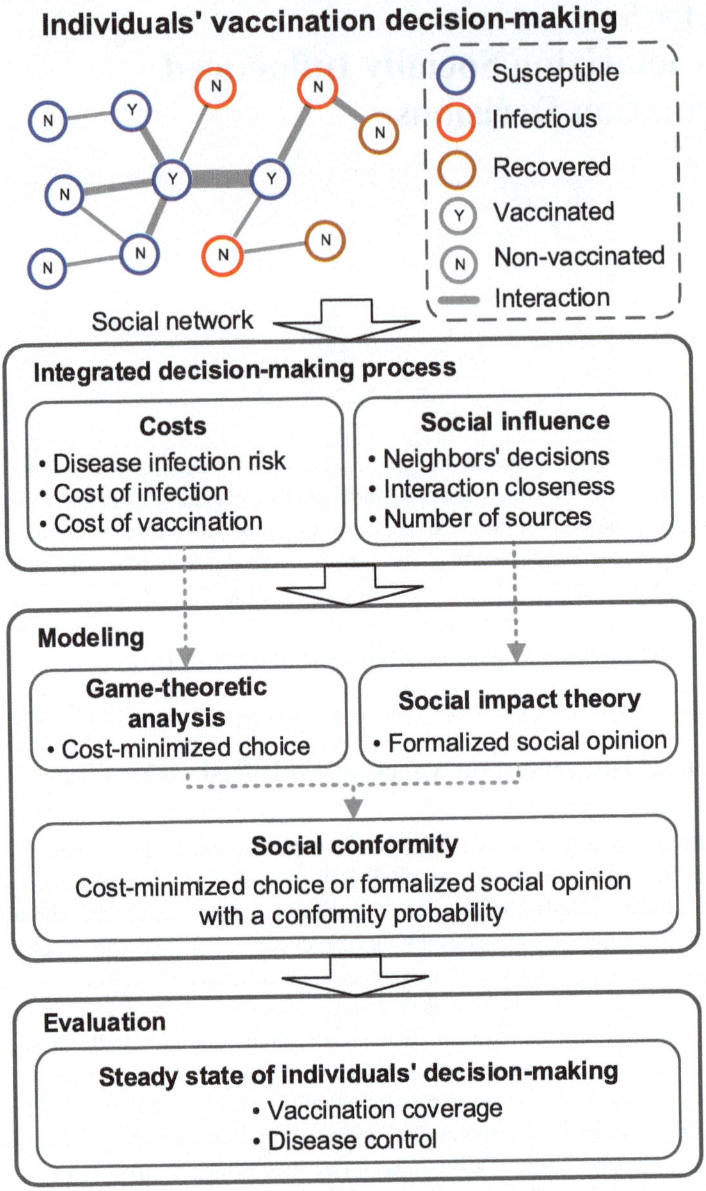

Fig. 5.1 An integrated decision process for characterizing individuals' voluntary vaccination. We extend the existing game-theoretic analysis by incorporating the effect of social influences. By doing so, we can investigate the steady state of individuals' decision-making and examine the effect of social influences on voluntary vaccination and, hence, the effectiveness of disease control

an extended SIT-based characterization of social influences with a game-theoretic analysis of cost minimization. In this model, we use conformity to describe the effect of social influences on vaccination decision-making, in terms of individuals' tendency to be affected by the social influences of others. Additionally, we represent individuals' interaction relationships with reference to a social network structure, in which individuals are heterogeneously interconnected with different numbers of connected neighbors and where there is variable social closeness in their interactions. We parameterize our model with an influenza-like disease and a real-world social network.

Through a series of simulations of voluntary vaccination, we examine the steady state of individuals' decision-making and evaluate the vaccination dynamics and the effect of disease control, in terms of vaccination coverage and the resulting infection rate, respectively. By doing so, we investigate the interplay of cost minimization and social influences in individuals' vaccination decision-making, and examine the effect of different levels of individuals' conformity on the strength of social influences. Furthermore, we provide a new modeling framework that incorporates the effect of social influences to investigate the effectiveness of voluntary vaccination for infectious disease control.

In addition to cost minimization, an individual may be affected by the decisions of others under the effect of social influences, and thus convert his/her cost-based choice to the social opinion of his/her neighbors, as illustrated in Fig. 5.2. According to SIT, the strength of such social influences is subject to the structure of individuals' interactions, e.g., the types of opinions (acceptance or rejection σ_i), interaction relationships (social closeness w_{ij}), and the number of opinion sources (the numbers of vaccinated and unvaccinated neighbors, N_i^{vac} and N_i^{non}, respectively). In our social network, for individual i, the strengths of social influences for two opposite opinions (vaccination acceptance and rejection), described by ι_i^{vac} and ι_i^{non}, can be accordingly computed as follows:

$$\iota_i^{vac} = \left(N_i^{vac}\right)^{1/2} \cdot \sum_{j \in N_i^{vac}} w_{ij}^2 \tag{5.1}$$

$$\iota_i^{non} = \left(N_i^{non}\right)^{1/2} \cdot \sum_{j \in N_i^{non}} w_{ij}^2 \tag{5.2}$$

We use $\tilde{\sigma_i}$ to denote the formalized social opinion resulting from the social influences of individual i's neighbors. As a modification of the standard SIT definition (where $\tilde{\sigma_i}$ corresponds to the opinion backed by stronger social influences), $\tilde{\sigma_i}$, being either acceptance or rejection of vaccination, will be determined by comparing the influences of the two opposite opinions. We let $\Delta\iota_i$ denote the discrepancy between ι_i^{vac} and ι_i^{non}. Then, we normalize $\Delta\iota_i$ as follows:

$$\Delta\iota_i = \frac{\iota_i^{vac} - \iota_i^{non}}{\iota_i^{vac} + \iota_i^{non}}. \tag{5.3}$$

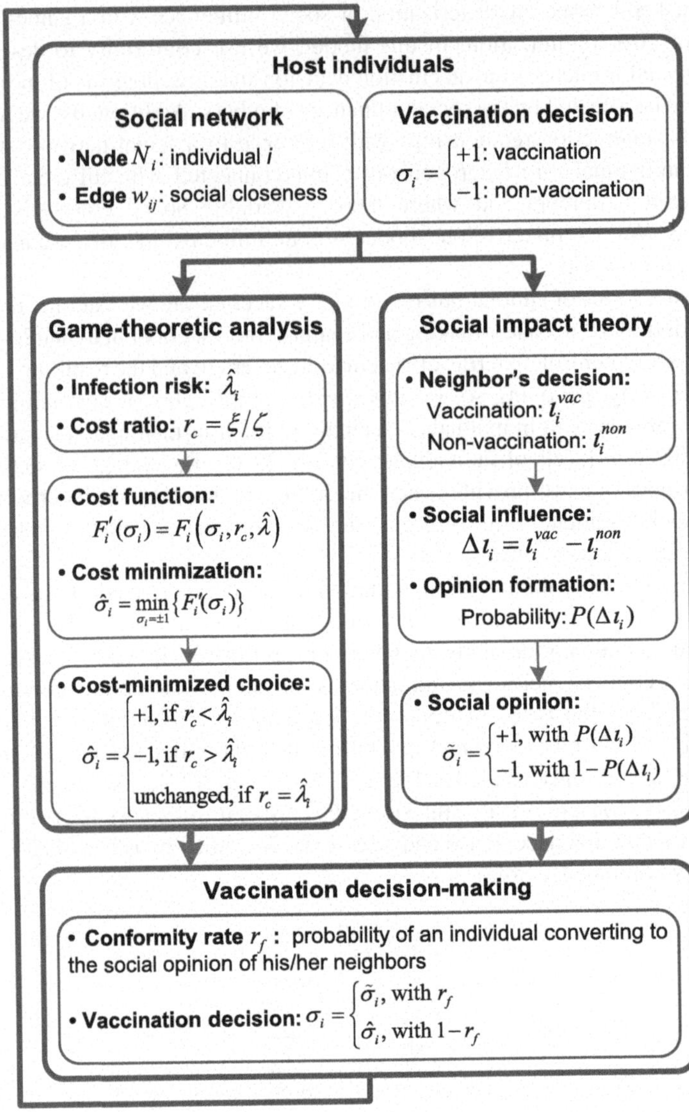

Fig. 5.2 A decision process for characterizing individuals' voluntary vaccination. We extend the existing decision analysis by incorporating the effect of social influences. A group of socially interconnected individuals can make decisions by evaluating the decisions of others. We utilize SIT to characterize the effect of social influences on individuals' decision-making with reference to their interaction relationships

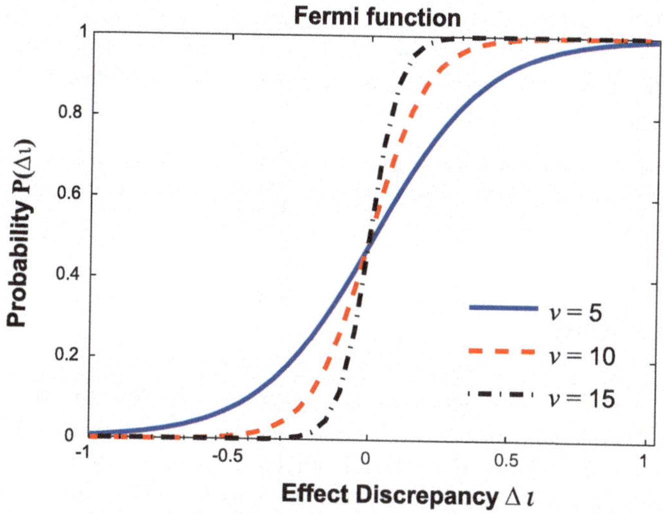

Fig. 5.3 Social opinion based on the influence of two opposite opinions. For individual i, $\tilde{\sigma}_i$ is formalized as acceptance of vaccination with the probability $P(\Delta\iota_i)$ or otherwise with the probability $1 - P(\Delta\iota_i)$. In the Fermi function, ν denotes individuals' responsiveness to the discrepancy $\Delta\iota_i$

Therefore, we can write $\tilde{\sigma}_i$ in the following form:

$$\tilde{\sigma}_i = \begin{cases} +1, & \text{with probability } P(\Delta\iota_i) \\ -1, & \text{with probability } 1 - P(\Delta\iota_i) \end{cases} \tag{5.4}$$

where $P(\Delta\iota_i)$ denotes the probability that social opinion $\Delta\iota_i$ is to accept vaccination, and $1 - P(\Delta\iota_i)$ to reject vaccination. Here $P(\Delta\iota_i)$ is computed from the Fermi function as follows:

$$P(\Delta\iota_i) = \frac{1}{1 + \exp(-\nu \cdot \Delta\iota_i)} \tag{5.5}$$

The Fermi function is a sigmoid function that has been widely used for describing individuals' behavioral changes as a response to the payoff discrepancy between two different choices. Here, ν describes individuals' responsiveness to the effect discrepancy, i.e., the difference in effect of the two opposite opinions. As shown in Fig. 5.3, a larger value of ν means that the choice backed by a stronger social influence will be more likely to dominate social opinion even if the effect discrepancy, $\Delta\iota_i$, is relatively small.

Next, we introduce a probability, r_f, termed the individual's conformity rate, that indicates the degree of individuals' tendency to adopt the social opinion of his/her interconnected neighbors, which corresponds to how likely individual i is to convert his/her cost-minimized choice ($\hat{\sigma}_i$) to the social influence-formalized opinion ($\tilde{\sigma}_i$).

Thus, $r_f = 0$ corresponds to the case of an absolute cost-based decision maker, whereas $r_f = 1.0$ indicates that the individual is an absolute social follower (i.e., ignoring his/her own cost evaluation). In other words, the final decision of individual i can be expressed as follows:

$$\sigma_i = \begin{cases} \tilde{\sigma}_i, & \text{with } r_f \\ \hat{\sigma}_i, & \text{with } 1 - r_f \end{cases} \tag{5.6}$$

5.2 Case Study

Further to the above description of the integrated decision-making model, we now show several simulation-based experiments based on the real-world epidemic scenario of the 2009 Hong Kong H1N1 influenza together with a social network with a total of 788 nodes and the average node degree of 35 (see also Sect. 4.3.1).

5.2.1 *Vaccination Coverage*

To evaluate the effect of individuals' vaccination decision-making on disease control, we further construct a standard SIR model to describe the threshold of vaccination coverage for mitigating an epidemic. The mathematical description of the threshold of vaccination coverage can be found in Sect. 3.1.1.

Based on our decision model, we conduct a series of simulations on vaccination dynamics to assess the vaccination coverage at the steady state of individuals' decision-making. As shown in Fig. 5.4, we investigate the interplay of cost minimization and the effect of social influences on individuals' vaccination decision-making with reference to three initial levels of vaccination willingness: 30%, 45%, and 60%. The level of vaccine uptake is subject to cost ratio r_c and individuals' initial level of vaccination willingness, where individuals' conformity rate r_f takes different values. Specifically, the simulation results in Fig. 5.4a–c show that when the effect of social influences is relatively weak (i.e., conformity rate r_f is relatively small), the cost of vaccination (cost ratio r_c) fundamentally determines the resulting vaccination coverage, in that increasing the cost of vaccination will lower individuals' vaccination willingness (at the steady state of individuals' decision-making). In our considered situation, the vaccination coverage is approximately 31% when cost ratio $r_c = 1.0$. Gradually, as r_c decreases and approaches 0, the vaccination coverage increases to 90%. Based on our model design, when an individual perceives that all of his/her interconnected neighbors have decided to be vaccinated, the individual will keep his/her previous choice of non-vaccination even if the cost of vaccination is zero, due to the consideration that disease transmission will no longer occur.

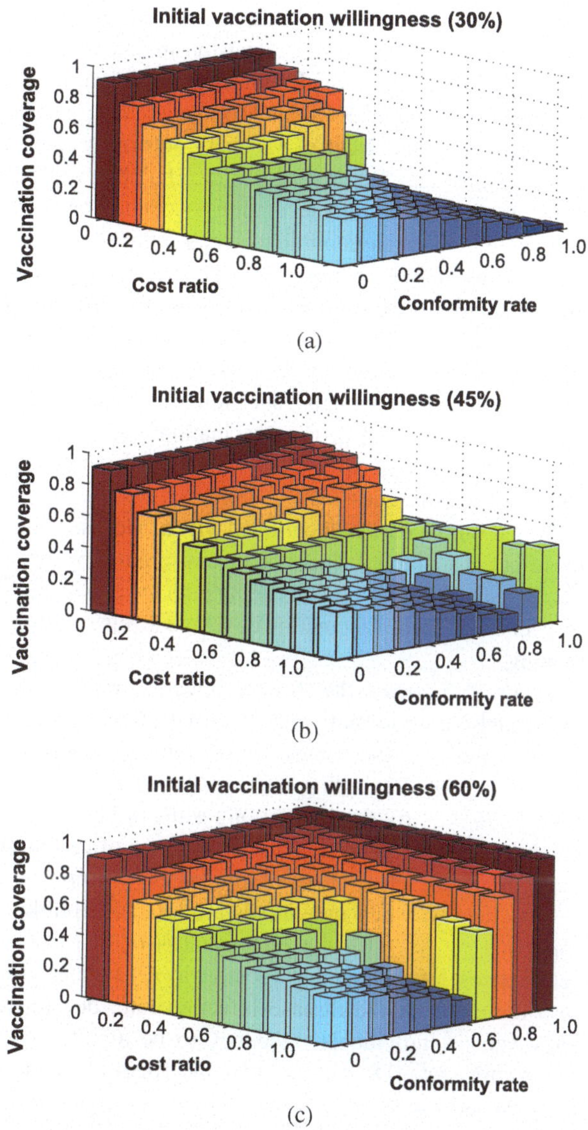

Fig. 5.4 Vaccination coverage at the steady state of individuals' decision-making. We investigate the interplay of cost minimization and the effect of social influences on individuals' vaccination decision-making, as measured by the resulting vaccination coverage, by varying the values of cost ratio r_c and conformity rate r_f between 0 and 1.0. Individuals' initial level of vaccination willingness is set as: (**a**) 30%; (**b**) 45%; and (**c**) 60%

Furthermore, we can observe that the strength of social influences (conformity rate r_f) moderates the aforementioned effect of cost ratio r_c on individuals' vaccination decisions. In the extreme case that individuals are purely cost-based decision makers ($r_f = 0$), the resulting vaccination coverage will be completely determined by the relative cost of vaccination (cost ratio r_c). Conversely, in the extreme case that individuals are absolute followers of social opinion ($r_f = 1.0$), the effect of social influences will promote an unvarying vaccination coverage, the level of which depends on individuals' initial level of willingness rather than the associated costs. In this simulation setup, when $r_f = 1.0$, the vaccination coverage at the steady state of decision-making converges to approximately 2% for individuals' vaccination willingness at the initial level of 30% (Fig. 5.4a), 50% at the level of 45% (Fig. 5.4b), and 97% at the level of 60% (Fig. 5.4c).

In addition, the effect of varying conformity rate r_f (individuals' tendency to be swayed by social opinions) can also be observed from the changing vaccination decisions depending on vaccination-associated costs (cost ratio r_c). When individuals become more susceptible to social influences (gradually increasing conformity rate r_f), as shown in Fig. 5.4a, the effect of such influences tends to increase the vaccination coverage when the cost of vaccination is low ($0 < r_c \leq 0.5$). Conversely, when the cost of vaccination is relatively high ($0.5 < r_c \leq 1.0$), the effect of social influences will reduce the resulting vaccination coverage at the steady state of individuals' decision-making. Furthermore, when the conformity rate r_f approaches 1.0, the vaccination coverage will decrease/increase sharply and finally converge to a fixed level that depends on individuals' initial level of vaccination willingness.

Based on the earlier-mentioned SIR model (see also Sect. 2.1.2), we investigate the effect of social influences on disease control by evaluating disease infection rates (the percentage of individuals being infected as a result of disease transmission) with respect to different vaccination coverage levels resulting from individuals' decision-making.

Figure 5.5 shows the disease infection rates with respect to the interplay of individuals' cost minimization and the effect of social influences on vaccination decision-making (the values of cost ratio r_c and conformity rate r_f each ranging from 0 to 1.0). Regarding our considered epidemic scenario (basic reproduction number $R_0 = 1.6$), the simulation results in Fig. 5.5a–c show that disease infection can be eliminated given a relatively low cost of vaccination (cost ratio $0 < r_c \leq 0.8$) and a moderate effect of social influences (conformity rate $0 < r_f \leq 0.6$).

Specifically, when individuals are less susceptible to social influences (conformity rate $r_f < 0.6$), the effectiveness of disease control is determined by the relative cost of vaccination (cost ratio r_c) in that a lower vaccination cost leads to a reduction in the disease infection rate, due to higher vaccination coverage. However, as individuals' tendency to be affected by social influences increases (conformity rate $0.8 \leq r_f < 1.0$), the effect of vaccination cost on disease control is weakened accordingly, while individuals' initial level of vaccination willingness matters more. In the extreme case of $r_f = 1.0$ (i.e., individuals are absolute followers of social influences), the disease infection rate increases to 46% for the initial level of vaccination willingness of 30%, as shown in Fig. 5.5a. If the initial level of vaccination willingness is set as 45%, as shown in Fig. 5.5b, the infection rate will be higher than in the situation with the initial level of 60%, as shown in Fig. 5.5c, where cost ratio $r_c > 0.8$ and conformity rate $0.2 \leq r_f < 0.8$.

In addition, we examine the steady-state vaccination coverage and the resulting infection rate corresponding to different initial levels of individuals' vaccination willingness prior to their decision-making, the results of which are shown in Fig. 5.6. We note that individuals' initial level of willingness affects the converged level

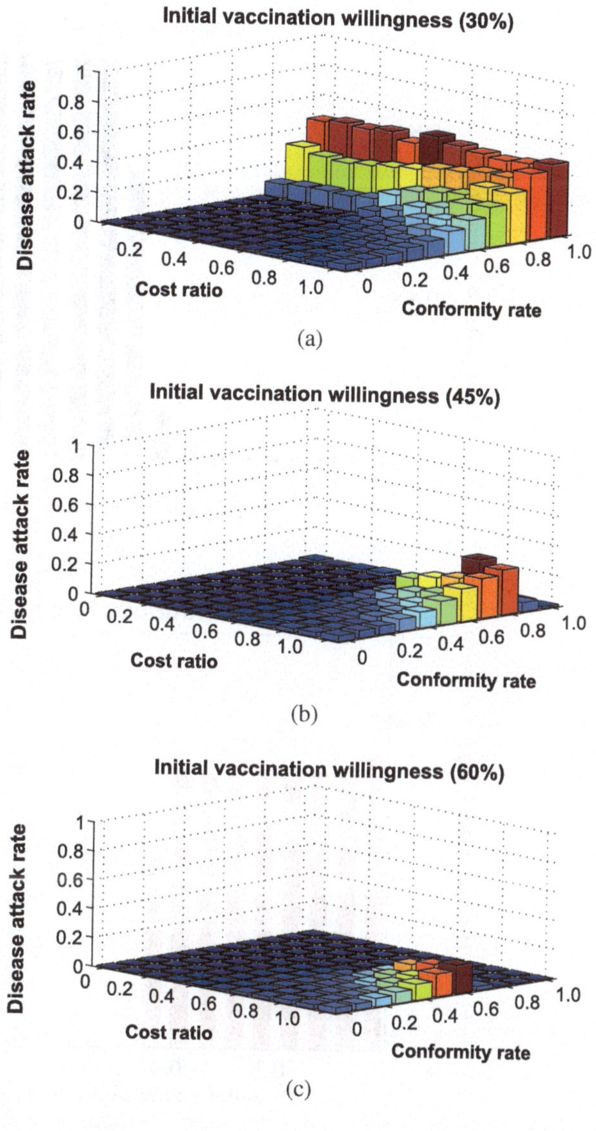

Fig. 5.5 Infection rates with respect to different levels of vaccination coverage resulting from individuals' decision-making. We investigate the interplay of cost minimization and the effect of social influences on the effectiveness of disease control, as measured by the percentage of individuals being infected as a result of disease transmission, through varying the values of cost ratio r_c and conformity rate r_f between 0 and 1.0. Individuals' initial level of vaccination willingness is set as: (**a**) 30%; (**b**) 45%; and (**c**) 60%

of the steady-state vaccination coverage as well as the effectiveness of disease control when individuals are absolute followers of social opinion (conformity rate $r_f \approx 1.0$). In our simulations, when the initial level of individuals' vaccination willingness is 30%, the converged steady-state vaccination coverage is approximately 2.4%, as shown in Fig. 5.6a. The vaccination coverage will reach 45% and 91% if the initial levels of vaccination willingness are 45% and 60%, respectively. In addition, we observe a critical phase transition in vaccination coverage when individuals' initial level of vaccination willingness is between 0.4 and 0.5, as shown

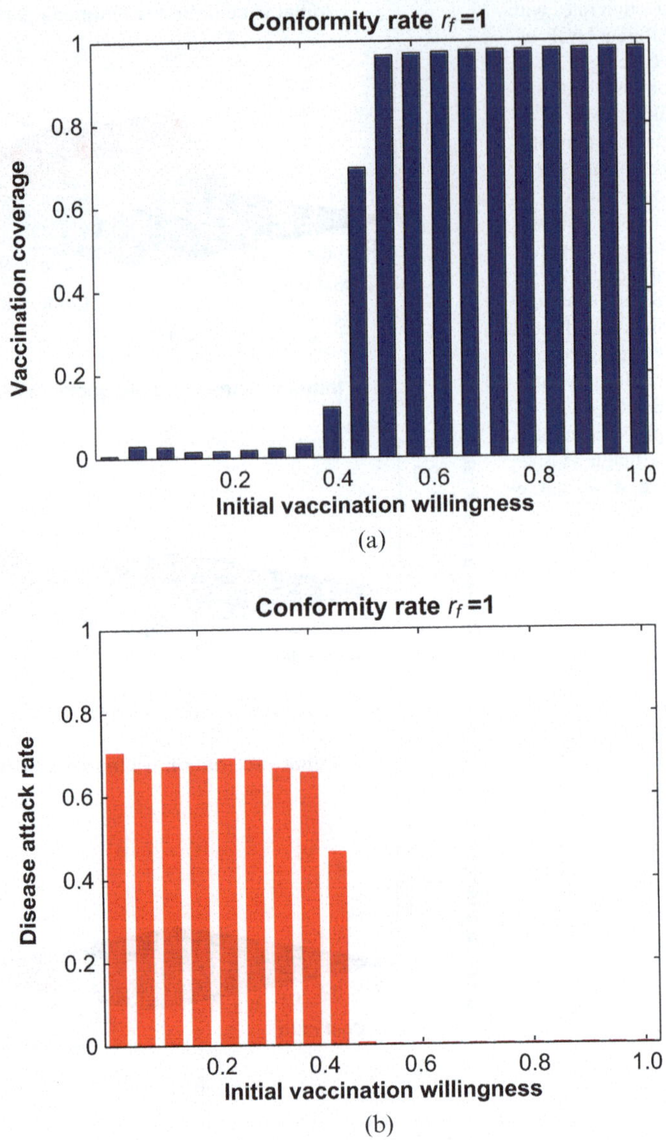

Fig. 5.6 The effect of individuals' initial levels of vaccination willingness when individuals are all social followers (conformity rate $r_f = 1.0$). (**a**) Vaccination coverage at the steady state of individuals' decision-making. (**b**) The resulting effects on epidemic control in terms of disease attack rate

in Fig. 5.6a. That is, in the situation of individuals being absolute social followers, there is a threshold value of individuals' initial level of vaccination willingness, which can be used to evaluate the effectiveness of a voluntary vaccination program for eliminating the epidemic, as in Fig. 5.6b.

Figure 5.7 shows the vaccination coverage at the steady state of individuals' decision-making with respect to different disease reproduction numbers. Here, we observe a similar effect of social influences in all three considered situations: the effect of social influences will increase the vaccination coverage when the relative cost of vaccination r_c is low (see Fig. 5.7a–c), reduce it when r_c is relatively high (see Fig. 5.7g–i), and bring the coverage to a certain level when individuals become strong followers of social influences (conformity rate r_f approaches 1.0). The simulation results further show that when the effect of social influences is relatively weak (conformity rate $0 < r_f \leq 0.6$), the vaccination coverage will increase in the case of relatively severe disease transmission, i.e., large reproduction numbers (e.g., $R_0 = 2.0$). However, if the effect of social influences is strengthened (conformity rate r_f approaches 1.0), the vaccination coverage at the steady state of individuals' decision-making is mainly determined by individuals' initial level of vaccination willingness, rather than the related costs and disease severity.

5.3 Further Remarks

The phenomenon of social influences, in which individuals' behaviors or opinions are affected by their social environment, has long been observed and studied. In the context of vaccination, social influences can affect individuals' vaccination decisions and, thus, the effectiveness of disease control in terms of the resulting population vaccination coverage. In this chapter, we addressed the effect of social influences on individuals' vaccination decision-making, vaccination coverage, and disease control. By parameterizing our model based on a real-world contact network and the 2009 Hong Kong H1N1 influenza epidemic, we carried out a series of simulations on individuals' voluntary vaccination. The simulation results confirmed that the relative cost of vaccination (cost ratio r_c) is one of the determining factors of the voluntary vaccination coverage. In our simulations, the cost ratio is particularly decisive if individuals are relatively indifferent to social influences (conformity rate r_f is relatively small). However, if individuals become more susceptible to social influences (r_f is large), such influences increase the vaccination coverage when the cost of vaccination is low and, conversely, reduce the vaccination coverage when the cost is high. In the extreme case where individuals are absolute social followers (conformity rate $r_f = 1.0$), the vaccination coverage at the steady state converges to a certain level that solely depends on individuals' initial level of vaccination willingness, and is independent of the vaccination-associated costs.

Several mathematical models have been proposed that utilize payoff-based methods to characterize individuals' vaccination decision-making based on their perception of the risks and benefits of vaccination. As an improvement over the

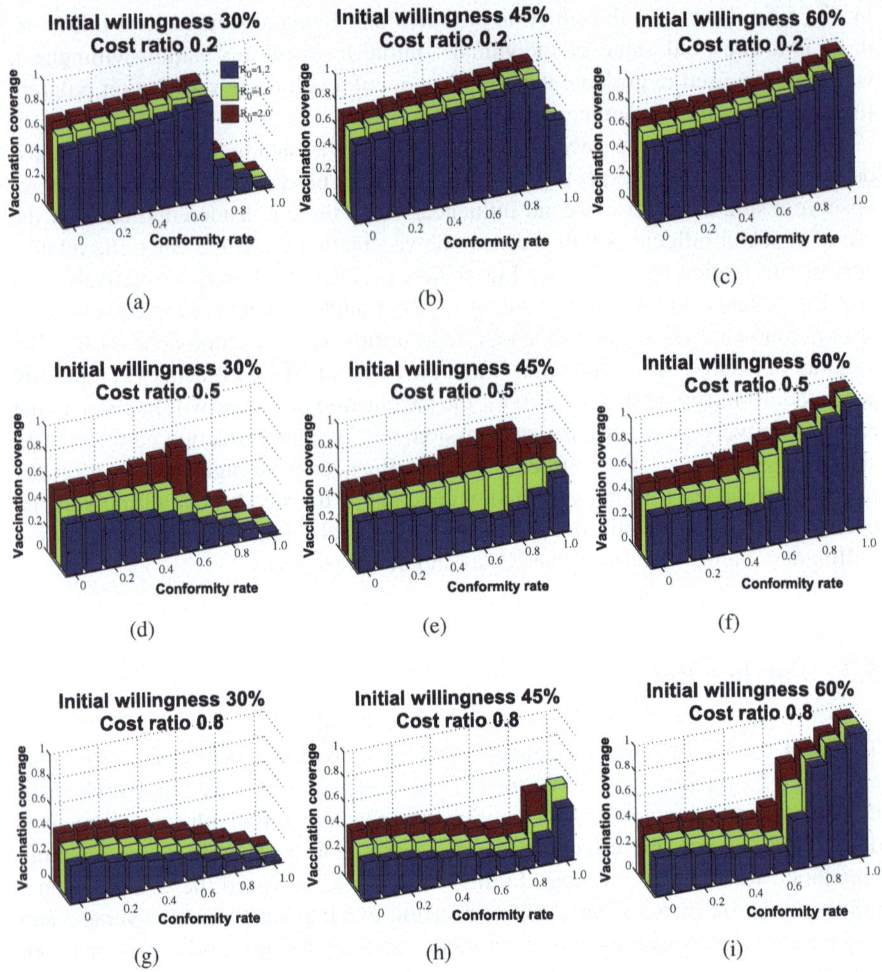

Fig. 5.7 Sensitivity analysis of the vaccination coverage at the steady state of individuals' decision-making, with respect to: (1) $R_0 = 1.2$; (2) $R_0 = 1.6$; and (3) $R_0 = 2.0$

existing models, we consider an individual's vaccination decision as an integrated process balancing his/her self-initiated cost minimization (e.g., behaviors that exploit the herd immunity effect) against the social influences of neighbors' decisions (social conformity). Our model introduces the parameter r_f (conformity rate) to modulate individuals' tendency toward the two decision-making mechanisms: an individual will either adopt his/her cost-minimized decision, or convert to the social opinion of his/her interconnected neighbors. Based on the existing studies that address individuals' vaccination decision-making as a process of independent opinion formation [164], we further take into account the heterogeneity of individuals' interaction relationships by an extended SIT-based characterization of the strength of social influences. Additionally, by incorporating the effect of social influences, we can investigate the effect of individuals' initial level of vaccination willingness on the vaccination coverage resulting from individuals' final decision-making.

By computationally characterizing the effect of social influences, our work has practical implications for understanding vaccination behaviors and improving the effectiveness of vaccination policies. In recent years, the rapidly increasing use of new communication tools e.g., Internet-based social media, has further amplified social influences. For example, both the efficacy and the adverse effects of vaccination are debated online, and opinions either for or against vaccination spread fast among users. We have identified that individuals' initial level of vaccination willingness is an important factor determining the final vaccination coverage, due to the effect of social influences (social conformity). Our results show that when conformity rate r_f approaches 1.0, the vaccination coverage at the steady state of individuals' decision-making will differ starkly, depending on different initial levels of vaccination willingness. Moreover, the empirical studies that have been performed to survey the determinants of individuals' vaccination decisions in a social environment readily provide us a practical means for measuring and evaluating conformity to social influences [165]. As has been shown in our study, individuals' vaccination decisions can be affected by both the associated costs and their conformity to social influences. Therefore, public health authorities must estimate the level of individuals' acceptance of a vaccine prior to the start of a voluntary vaccination program, as well as rapidly assess and enhance the effectiveness of their adopted vaccination policies, e.g., provide certain financial subsidies to reduce the cost of vaccination.

So far, our study has provided a modeling framework for incorporating the effect of social influences on individuals' decision-making and disease control. We note that the results demonstrated in this chapter may be reliant on the chosen social network, which was constructed using data from students' interactions within an American high school. In our model, the social influences account only for the localized interactions between an individual and his/her interconnected neighbors. Additionally, by utilizing an SIT-based characterization of social effect, we implicitly assume that individuals are passive recipients of social influences and do not actively shape them.

5.4 Summary

In this chapter, we further addressed the characterization of individuals' voluntary vaccination rate. To achieve this, we considered a group of individuals making their decisions with the aim of minimizing the associated costs while evaluating the decisions of others. In this regard, we introduced an integrated process of vaccination decision-making that incorporates both the cost analysis of vaccination and the effect of social influences. In addition, we applied SIT to characterize the effect of social influences on individuals' changing vaccination decisions corresponding to their interaction relationships. By doing so, we conducted an in-depth study of vaccination decision-making that integrated an extended SIT-based characterization of social influences with a game-theoretic analysis of cost minimization. We adopted individuals' social network to capture the structure of their interaction relationships and carried out a series of simulations on voluntary vaccination against an influenza-like disease. By evaluating the resulting coverage of voluntary vaccination, we evaluated the effect of social influences on individuals' decisions and, in turn, on the effectiveness of disease control (vaccination coverage), subject to three determinants, as follows: (1) the relative cost of the decision to vaccinate, i.e., cost ratio r_c; (2) individuals' susceptibility to social influences, i.e., conformity rate r_f; and (3) individuals' initial level of vaccination willingness. Based on the results in this chapter, we conclude that social influences will have an effect on the ability to control an infectious disease by indirectly affecting the vaccination coverage. In this respect, our work can provide a means to model the effect of social influences and estimate the effectiveness of a voluntary vaccination program.

Chapter 6
Understanding the Effect of Social Media

It has long been observed that the spread of awareness about an epidemic affects individuals' behaviors or opinions. In the context of an emerging infectious disease, it would be unrealistic for individuals to have prior knowledge about a newly developed vaccine when deciding whether to be vaccinated. In this case, the spread of awareness on social media about both disease severity and vaccine safety will affect individuals' subjective perception of vaccination and, hence, substantially influence their decisions.

In this chapter, we address the further development of decision models enabling us to characterize individual-level voluntary vaccination; the models should incorporate the factor of subjective perception, which is partly shaped by social media [54]. In addition, we examine how such individuals' vaccination decisions affect efforts to ensure adequate vaccination coverage for infectious disease control.

6.1 Modeling Subjective Perception

In the preceding two chapters, voluntary vaccination decision-making was modeled from the perspective of rational individuals who assess the socio-economic costs of disease infection and vaccination (see also Sect. 4.1). In this regard, game-theoretic analysis was used to describe individuals' vaccination decisions by evaluating their personal optimized payoffs based on the perceived risks and benefits of vaccination. Payoff-based decision-making models make the assumption that individuals have prior knowledge about the disease, the vaccine, and the associated costs, which may be valid for routine vaccination programs against seasonal infectious diseases (e.g., measles and chickenpox). However, for a newly developed vaccine against an emerging infectious disease (e.g., the 2009 H1N1 influenza), there will be a lack of such knowledge. Thus, the assumption of individuals' rational decision-making may not hold.

J. Liu, S. Xia, *Computational Epidemiology*, Health Information Science,
https://doi.org/10.1007/978-3-030-52109-7_6

As we have discussed, individuals' vaccination decisions are influenced by their perception of disease severity and vaccine safety. Such perception may be informed by news of recent events relayed by others [166]. That is to say, individuals' perception can be mediated by the spread of awareness in a host population, which can potentially alter their vaccination decisions and, hence, affect disease transmission and the effectiveness of vaccination programs.

In recent years, rapidly emerging social media, such as Facebook [167], Twitter [168], and YouTube [169, 170], have served as new channels for the spread of public health information. Specifically, the efficacy of vaccines is widely debated among online social communities; reports of the adverse events of vaccination are shared across the Internet, and opinions either for or against vaccination can be effectively "transmitted" from person to person. In this case, the spread of awareness about an infectious disease and a vaccine can be as fast as the spread of the disease itself, so that individuals will respond immediately to negative events as they occur. Therefore, the dynamics of individuals' vaccination will be tightly coupled with that of disease transmission. The interplay between the two dynamic processes can have significant consequences for the resulting vaccination coverage for infectious disease control.

In this chapter, we study individuals' vaccination decisions under the influence of the spreading awareness of disease- and vaccine-related negative events during an epidemic. As illustrated in Fig. 6.1, we consider a group of individuals who decide whether or not to take a vaccine based on their perceptions of disease severity and vaccine safety. Specifically, individuals interact through their social relationships (e.g., friendships on Facebook and follower relationships on Twitter). In such a structured host population (represented by a social network), awareness about negative events can spread from person to person and substantially affect individuals' perceptions of disease severity and vaccine safety. On the one hand, reported cases of severe infection will cause individuals to perceive the disease as more serious and, hence, increase their tendency to be vaccinated. On the other hand, reported events of the vaccine's adverse effects will weaken the public confidence in vaccine safety, leading to lower acceptance of vaccination.

In this case, we develop a novel decision model for characterizing individuals' voluntary vaccination rate, in which we suppose that an individual will voluntarily decide to accept or reject the vaccine based on his/her beliefs on whether or not vaccination is acceptable. If the individual is not sufficiently confident in his/her view for or against vaccination, he/she will not make any firm decision but wait to see future developments. This situation in which an individual makes no firm decision can be considered as a state of "yet to decide", due to uncertainty. In this regard, we introduce three belief variables in the form of *Yes*, *No*, and *No decision* to characterize the possible decision of an individual to accept or reject the vaccine, or to have not yet decided. Due to the spread of awareness, the individual will update his/her beliefs about vaccination by collecting information from his/her social neighbors, which may either reinforce his/her own perception, or bring about conflicting perception.

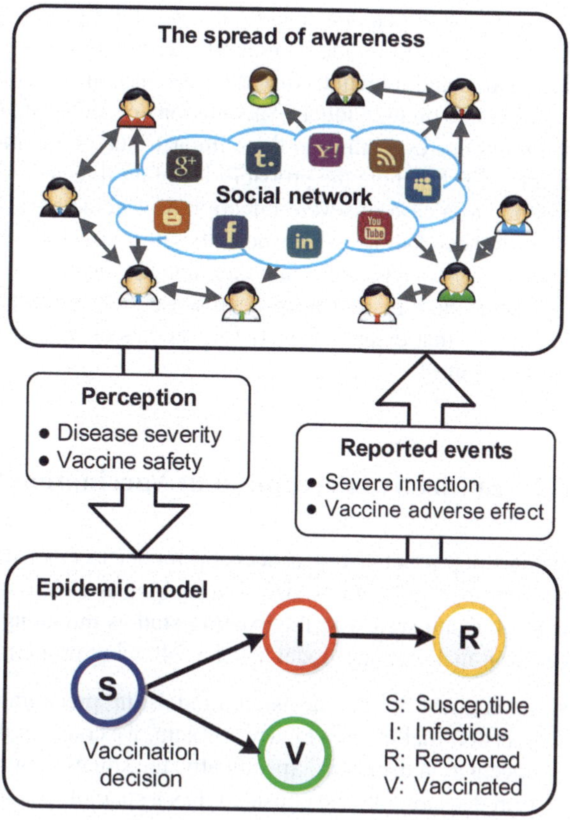

Fig. 6.1 A schematic illustration of the spread of awareness affecting individuals' vaccination decision-making. We consider a group of individuals whose vaccination decisions depend on their perception of disease severity and vaccine safety. Awareness about negative events can spread from person to person through their interaction relationships, which will substantially affect their perceptions of disease severity and vaccine safety and, thus, their vaccination decisions

To characterize individuals' evolving belief values based on the spread of awareness and their subsequent vaccination decision-making in the presence of uncertainty, we develop a new belief-based decision model by extending and utilizing the framework of the Dempster-Shafer Theory (DST) [171]. DST, also known as the theory of beliefs, was originally proposed as a generalization of the Bayesian theory of subjective probability to characterize how individuals update their beliefs by combining new pieces of evidence from multiple sources in the presence of uncertainty [172]. In our DST-based model, individuals can update their beliefs (with respect to perceived disease severity and vaccine safety) by combining the new evidence collected from social neighbors (obtained awareness about negative events). Furthermore, we extend the conventional DST framework by incorporating the effect of awareness spreading and fading, in which the awareness about a negative event (a piece of new evidence) will ripple through individuals' social networks, but the certainty of belief regarding that event (the belief values of both *Yes* and *No*) will decay gradually. In this regard, individuals' vaccination decision-making is modeled as a process affected by the spread of awareness about negative events and the subsequent updating of their beliefs.

We parameterize our model with an influenza-like disease and a social network from a real-world online community. Through a series of simulations of voluntary vaccination and infectious disease transmission, we evaluate the effect of the spread of awareness on individuals' vaccination decisions and the consequences for disease dynamics corresponding to three impact factors: (1) the reporting rates of negative events, which denote the probabilities of an infected or vaccinated individual being reported as a case of severe infection or vaccine adverse effects; (2) the coefficient of awareness fading, which describes the effect of certainty decay as awareness spreads from one person to another; and (3) the disease reproduction number, which represents the infectiousness of a disease. Furthermore, we provide a new modeling framework that extends the existing studies on voluntary vaccination for infectious disease control.

6.2 Subjective Perception in Vaccination Decision-Making

We develop a belief-based decision model to evaluate the effect of the spread of awareness on individuals' decision-making as well as on the effectiveness of disease control. Compared with the existing studies modeling individual-level vaccination decision-making, our model has the following properties:

- In contrast to the decision models in preceding chapters, which represent decision-making as a binary problem, we consider the uncertainty in individuals' vaccination decisions. Specifically, the situation in which an individual makes no firm decision may be considered as a state of "yet to decide", due to uncertainty. In this regard, we introduce three belief variables to characterize the possible decisions of an individual to accept or reject the vaccine, or to have not yet decided.
- We further consider the dependence of individuals' decisions on their subjective perception about the acceptability of vaccination. Moreover, awareness about disease severity and vaccine safety can spread from person to person and substantially affect personal perceptions of vaccination.
- To characterize the spread of awareness, we utilize real-world online social networks to characterize the structure of individuals' interactions. By doing so, we further utilize and extend DST to characterize individuals' belief updating and their decision-making, having incorporated the awareness obtained from their socially interconnected neighbors.

We consider a voluntary vaccination program for preventing the outbreak of an emerging infectious disease, e.g., 2009 H1N1 influenza, in which individuals can decide whether or not to take the vaccine based on their perception of disease severity and vaccine safety. It is assumed that individuals do not possess any prior knowledge about the disease or vaccine, while they can receive information about related events (reported cases of severe infection and vaccine adverse effects). In this case, awareness about these negative events will spread among the host individuals, rippling through their interaction relationships, which will in turn affect their vaccination decisions.

For such a situation, we construct a new individual-level model to characterize vaccination decision-making. At the same time, we use an epidemic model to describe the dynamics of disease transmission as a result of individuals' voluntary vaccination. Based on our model, we investigate how the spread of awareness affects the changes of individuals' vaccination decisions during an infectious disease outbreak.

6.2.1 Dempster-Shafer Theory (DST)

We represent an individual's willingness to accept or reject vaccination by using a set of belief variables. Based on the Dempster-Shafer Theory (DST), we suppose that the decision on whether or not to be vaccinated is a binary problem, which is represented as $\Theta = \{Yes, \ No\}$. Here Θ is the frame of discernment for the vaccination decisions (a universal set). Individuals' possible vaccination-decision responses can be modeled as subsets of Θ, i.e., belonging to a power set, $2^{\Theta} = \{\phi, \{Yes\}, \{No\}, \Theta\}$. Next, we use a function $m(\cdot)$ to assign a belief mass (probability) to each element of the power set 2^{Θ}, which is called the basic probability assignment (BPA). The mass $m(A)$ $(A \in 2^{\Theta})$ denotes the proportion of support for the particular subset A based on the current available evidence or knowledge. The BPA has the following two properties: (1) the mass of empty set ϕ is zero; and (2) the masses of the power set sum to one:

$$\begin{aligned} &m : 2^{\Theta} \rightarrow [0, 1] \\ &m(\phi) = 0 \\ &\sum_{A \subseteq 2^{\Theta}} m(A) = 1 \end{aligned} \tag{6.1}$$

Accordingly, the belief functions for an individual's vaccination-decision responses can be expressed as follows:

$$\begin{aligned} &m(\phi) = 0 \\ &m(Yes) \in [0, 1] \\ &m(No) \in [0, 1] \\ &m(\Theta) = 1 - m(Yes) - m(No) \end{aligned} \tag{6.2}$$

where $m(Yes)$ describes an individual's belief that he/she should be vaccinated to prevent infection, $m(No)$ represents the belief that he/she should reject vaccination having considered the potential adverse effects, and $m(\Theta)$ denotes the belief that he/she is yet to decide whether or not to be vaccinated (due to uncertainty about the disease and vaccine). Based on the above formulation, an individual will decide to be vaccinated with the probability of $m(Yes)$, reject the vaccine with the probability of $m(No)$, and have no firm decision with the probability of $m(\Theta)$. In addition, we assume individuals who decide to be vaccinated will be vaccinated immediately and, therefore, will either be successfully immunized or suffer from vaccine adverse effects. Those with no decision will review their decision state in the next time step.

6.2.2 Spread of Social Awareness

During the spread of an emerging infectious disease and the implementation of a vaccination program, individuals' perceptions of disease severity and vaccine safety will be affected by their awareness of negative events caused by both the disease and vaccine. On the one hand, a case of severe infection can be naturally regarded as evidence that an individual should be vaccinated. On the other hand, news of the vaccine's adverse effects can be viewed as evidence that vaccination should be rejected.

Here, we use a belief value m^e to characterize a piece of evidence of a reported negative event that spreads through the host population. We suppose that a group of individuals are interconnected through their social network, denoted by $G = \langle V,\ L\rangle$, where $V = \{v_1, v_2, \cdots, v_N\}$ is the set of nodes (individuals), and $L = \{\langle v_i, v_j\rangle \mid 1 \leq i, j \leq N,\ i \neq j\}$ is the set of links (social interaction relationships). N is the total number of individuals. During an epidemic, each reported negative event will be treated as a piece of triggering evidence with a belief value of $m^e = \{m^e(Yes),\ m^e(No)\}$, where $m^e = \{1.0,\ 0\}$ and $m^e = \{0,\ 1.0\}$ for a reported case of severe infection and vaccine adverse effects, respectively.

Individuals can learn of emerging evidence by interacting with their social neighbors, update their belief values accordingly, and further spread the news to others through their social networks. However, the certainty of belief regarding a piece of evidence will decay as it is transmitted from person to person, which is referred to as awareness fading. Here, we introduce a fading coefficient, f, to indicate how fast the certainty will decay when a piece of evidence is transmitted between two individuals. A larger value of f corresponds to a faster decay (awareness fading). Therefore, the evidence that is transmitted (i.e., the spread of awareness) from individual j to his/her socially interconnected neighbor i can be computed as follows:

$$\begin{aligned}
m_i^e(Yes) &= (1 - f) \cdot m_j^e(Yes) \\
m_i^e(No) &= (1 - f) \cdot m_j^e(No) \\
m_i^e(\Theta) &= m_j^e(\Theta) + f \cdot \left(m_j^e(Yes) + m_j^e(No)\right)
\end{aligned} \tag{6.3}$$

In the course of disease transmission and vaccine implementation, newly reported negative events will constitute new sources of evidence at different time steps. The spread of awareness about the events will cause an individual to continuously update his/her perception of the disease and vaccine and then make his/her vaccination decision. Based on the obtained awareness, the individual will update his/her belief values (denoted by m_i') by combining the present belief values (denoted by m_i) with the newly received evidence m_i^e. This can be expressed in the following form (with $\oplus$ denoting the combination operation):

$$m_i' = m_i \oplus m_i^e \tag{6.4}$$

Specifically, based on the assumption that the multiple sources are independent, the belief-value update corresponding to the extended Dempster rule of combination is as follows:

$$m_i{}'(S_A) = \frac{\sum\limits_{S_B \cap S_C = S_A} m_i(S_B) \cdot m_i^e(S_C)}{1 - \sum\limits_{S_B \cap S_C = \phi} m_i(S_B) \cdot m_i^e(S_C)}, \quad \text{and} \quad S_A, S_B, S_C \in 2^{\Theta} \tag{6.5}$$

where $\sum\limits_{S_B \cap S_C = \phi} m_i(S_B) \cdot m_i^e(S_C)$ represents the basic belief mass associated with the conflict between present beliefs and the newly received evidence. In the Dempster combination rule, the denominator, $1 - \sum\limits_{S_B \cap S_C = \phi} m_i(S_B) \cdot m_i^e(S_C)$, is a normalization factor, which attributes the conflict probability mass to the universal set $m(\Theta)$.

By doing so, we utilize DST to develop a belief-based decision model for characterizing individuals' vaccination decisions in the presence of uncertainty. Furthermore, we extend the classical DST framework by incorporating the spread of awareness in a structured host population, in which the certainty regarding a piece of evidence will decay as it is transmitted from person to person.

For illustration, Fig. 6.2 shows individuals' belief-value updates calculated with respect to the spread of awareness about two independently reported negative events on a synthetic lattice network, in which the fading coefficients are set as: 0.1 (Fig. 6.2a) and 0.2 (Fig. 6.2b), respectively.

6.3 Case Study

We carry out a series of simulations to examine the coupled dynamics of individuals' voluntary vaccination and disease spread based on the real-world 2009 Hong Kong H1N1 influenza epidemic.

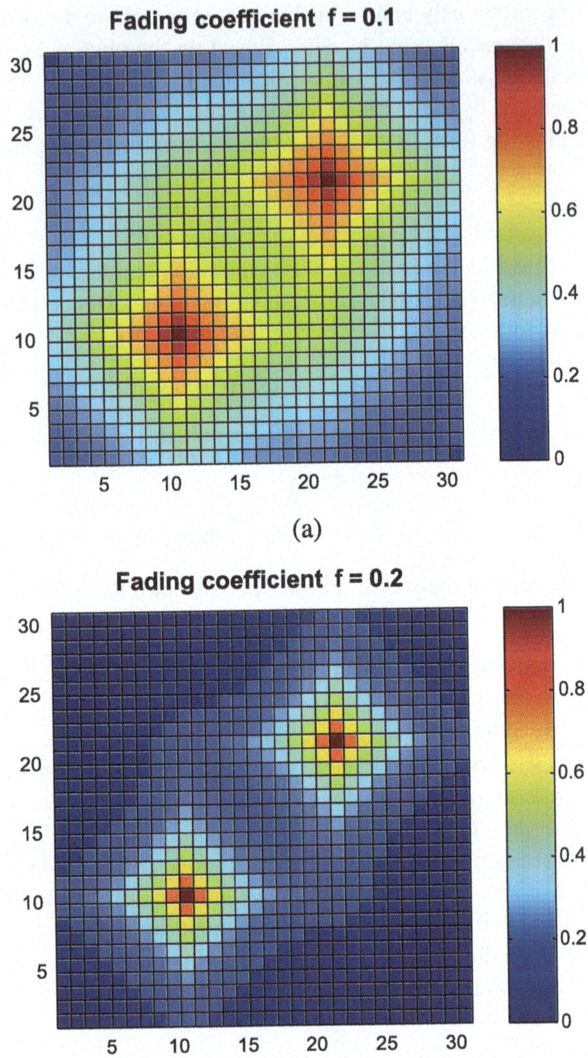

Fig. 6.2 Illustrations of awareness spreading and fading on a synthetic lattice network (30 × 30 nodes). Individuals can interact with their socially interconnected neighbors. The awareness of two reported negative events will spread independently in the structured host population, and affect individuals' belief values. Here, the color of a cell denotes an individual's belief value in terms of the probability mass ($m_i \in (0, 1.0)$). At the source of a reported negative event, the probability mass is set as $m_i = 1.0$. Parameter f denotes the coefficient of awareness fading. We demonstrate the effects of awareness fading corresponding to two coefficient values: (**a**) $f = 0.1$ and (**b**) $f = 0.2$

6.3.1 *Vaccination Decision-Making in an Online Community*

We construct an infectious disease model to characterize the spread of an emerging infectious disease in a host population, in which negative events (e.g., severe infections and vaccine adverse effects) will be reported and, hence, awareness of them will spread. In doing so, we adopt a standard SIR model, as mathematically described in Sect. 2.1.2. In addition, for the sake of example, we assume that a vaccine is available and adequately supplied at the time of disease onset. Only susceptible individuals can decide whether and when to be vaccinated. Once an individual is vaccinated, it is assumed that he/she will be completely immunized and move from the susceptible compartment to the recovered/immunized compartment.

For our simulations, we calibrate the parameters in our epidemic model with reference to the 2009 Hong Kong H1N1 influenza epidemic (see also Sect. 2.3.1). During that outbreak, there were more than 36,000 laboratory-confirmed cases (as of September 2010), among which approximately 290 were identified as severe (i.e., the reporting rate of severe infections was estimated as 0.805%). The Human Swine Influenza Vaccination Programme (HSIVP) was launched on December 1, 2009. The numbers of vaccinated individuals and reported negative events since that date are shown in Fig. 6.3. As of March 13, 2010, more than 180,000 doses of HSI vaccine were administered to persons of various groups (Fig. 6.3a) [173]. Throughout HSIVP as a whole, a total of 34 cases of adverse events following immunization (AEFI) were reported (Fig. 6.3b). The rate of AEFI was evaluated as 17.8 per 100,000 vaccinated individuals (i.e., the reporting rate of vaccine adverse effects was estimated as 0.0178%) [95].

We further construct a social network to characterize individuals' interaction relationships based on the data of a Facebook-like online community [174], in which registered users can communicate through personal blogs and forum postings. In this network, there are in total 1899 nodes and 13, 838 undirected links among them. As shown in the partial network snapshot of Fig. 6.4, the nodes denote the registered users and the links among them represent their interaction relationships, in terms of sending and receiving at least one message. Based on this network structure, we carry out a series of Monte Carlo simulations and experimentally examine the above-mentioned belief-based characterization of individuals' vaccination decision-making.

The simulation results in Fig. 6.5 show the dynamics of disease transmission and individuals' voluntary vaccination for the first 50 days. Figure 6.5a indicates the dynamics of disease transmission with the impact of individuals' voluntary vaccination decision-making. Figure 6.5b presents the distributions of simulated vaccine-induced negative events and severe disease infections over time. Figure 6.5c shows the changes of individuals' belief values in the three states. In this case, we reproduce the observed patterns of the real vaccination program, as shown in Fig. 6.5d, in terms of daily number of vaccinated individuals. We observe that the number of vaccinated individuals increases steadily in the earlier days of the vaccination program, as individuals' uncertainty about vaccination decreases. However, the reported cases of vaccine adverse effects significantly strengthen individuals' belief that non-vaccination is preferable, which causes the daily number of vaccinated individuals to fall sharply when this belief peaks on day 12.

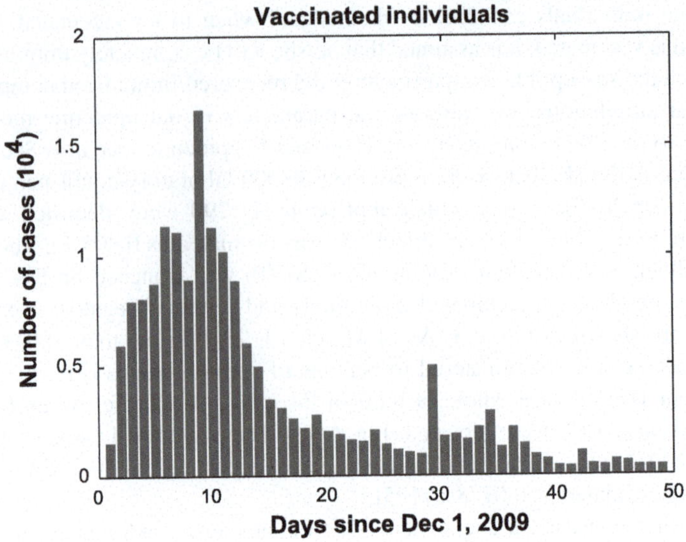

(a)

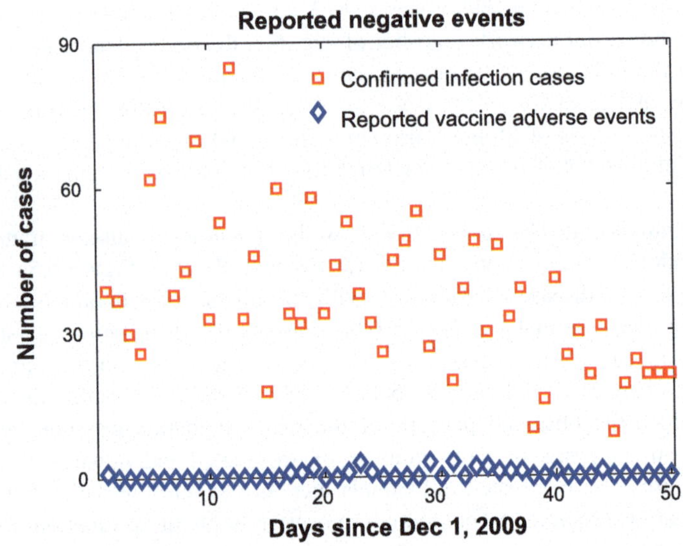

(b)

Fig. 6.3 The Human Swine Influenza Vaccination Programme (HSIVP) in Hong Kong. This program was launched on December 1, 2009. As of March 13, 2010, more than 180,000 doses of HSI vaccine were administered to persons of various groups. The rate of adverse events following immunization (AEFI) was evaluated as 17.8 per 100,000 vaccinated individuals, i.e., the reporting rate of AEFI was estimated as 0.0178%. (**a**) The daily number of vaccinated individuals since the beginning of HSIVP. (**b**) The reported cases of infection and vaccine adverse effects

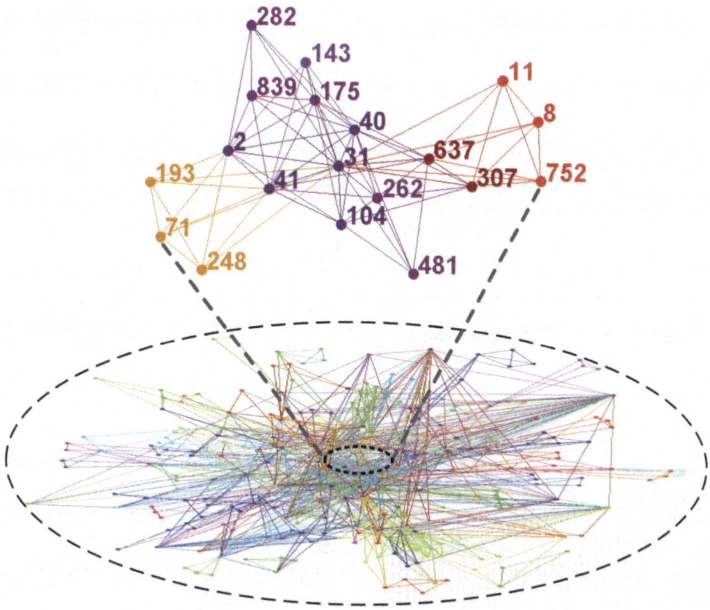

Fig. 6.4 A partial snapshot of individuals' social network. We utilize a network structure to represent individuals' interaction relationships, based on the data of a Facebook-like online community. In this network, the nodes denote individuals and the links represent their interactions in terms of sending and receiving messages

6.3.2 *Interplay of Two Dynamics*

We run our model with the aforementioned parameterization under various settings to reveal the interplay between the dynamics of disease transmission and individuals' vaccination. In doing so, we investigate the effect of the spread of awareness of disease severity and vaccine safety in a host population by investigating various settings of the reporting rates of negative events ϵ and κ, the coefficient of awareness fading f, and the disease basic reproduction number R_0.

As shown in Fig. 6.6, we investigate the reporting rates for the negative events of severe infections (ϵ) and vaccine adverse effects (κ) with respect to two levels: 1% and 0.1%. Here, we set $\epsilon = 0.01$ and $\kappa = 0.001$ for the situation of "disease scare", and similarity, $\epsilon = 0.001$ and $\kappa = 0.01$ for the situation of "vaccine scare". A higher reporting rate of severe disease infections will prompt individuals to be vaccinated (Fig. 6.6a, dash curve), which will in turn reduce disease transmission (Fig. 6.6b, dash curve). Moreover, vaccination in the early stage will be more effective than in the later stages. We can observe that when $\epsilon = 0.01$, the difference in the number of vaccinated individuals between the situations of $\kappa = 0.001$ and $\kappa = 0.01$ (Fig. 6.6a, dashed curve and solid curve, respectively) is relatively small in the early stage of disease transmission (before day 10). After that, the vaccination dynamics when

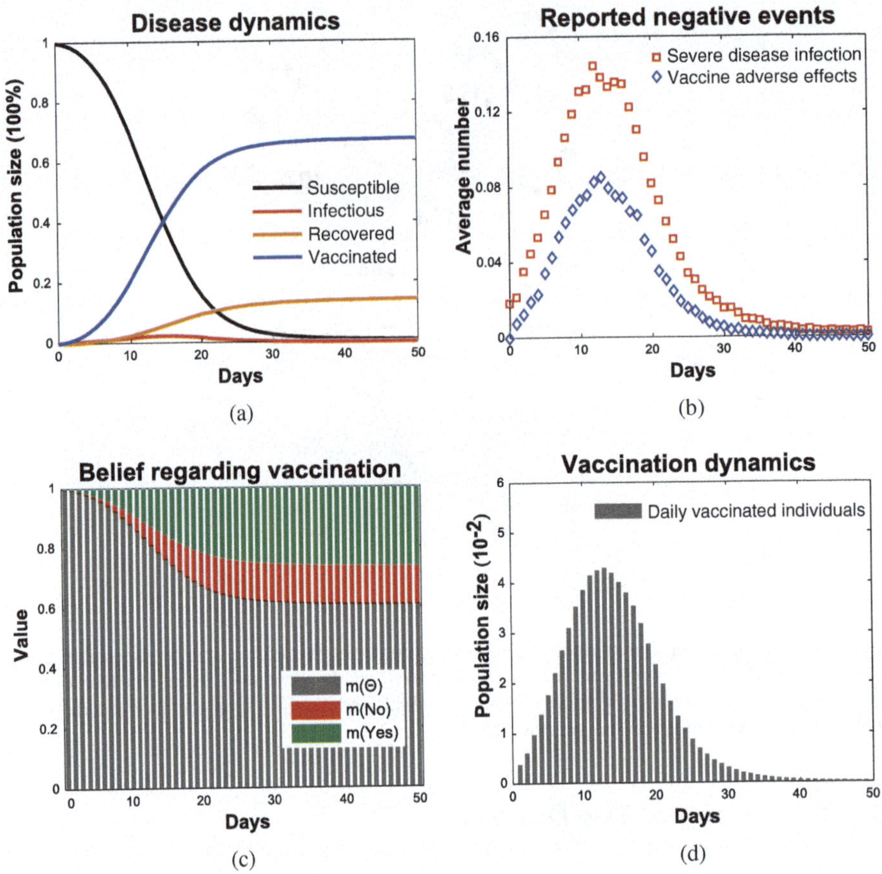

Fig. 6.5 Monte Carlo simulations of disease transmission and voluntary vaccination. (**a**) The dynamics of disease transmission in terms of the sizes of susceptible, infectious, recovered, and vaccinated populations. (**b**) The reported cases of severe disease infection and vaccine adverse effects. (**c**) The average belief values about vaccination in a host population. (**d**) The dynamics of voluntary vaccination in terms of the daily number of vaccinated individuals

$\epsilon = 0.01$ and $\kappa = 0.001$ will peak on day 15 with more than 4% individuals being vaccinated, while that of $\epsilon = 0.01$ and $\kappa = 0.01$ will peak on day 11 at 2%. Accordingly, we can observe that the disease dynamics in the situations of $\epsilon = 0.01, \kappa = 0.001$ and $\epsilon = 0.01, \kappa = 0.01$ (Fig. 6.6b, dashed curve and solid curve, respectively) have a similar incidence rate at the peak of disease infection, but the durations of the disease transmission period are different.

In addition, we investigate the effect of awareness fading corresponding to different fading coefficients f, the results of which are shown in Fig. 6.7. We note that awareness fading can affect the dynamics of individuals' vaccination in terms of the number of vaccinated individuals and the time of individuals' being vaccinated. In our simulation, when the fading coefficient $f = 0.1$, the daily

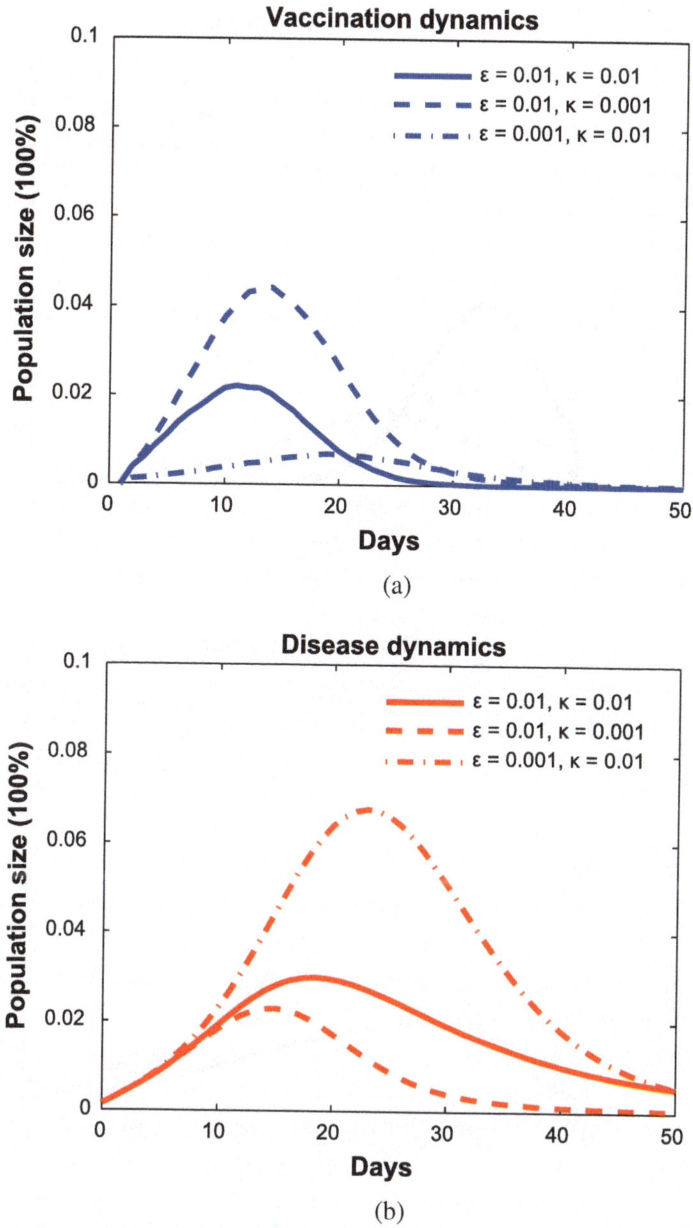

Fig. 6.6 The effect of reporting rates of negative events (severe disease infections ϵ and vaccine adverse effects κ). (**a**) The dynamics of voluntary vaccination (daily number of vaccinated individuals). (**b**) The dynamics of disease transmission (daily number of infectious individuals)

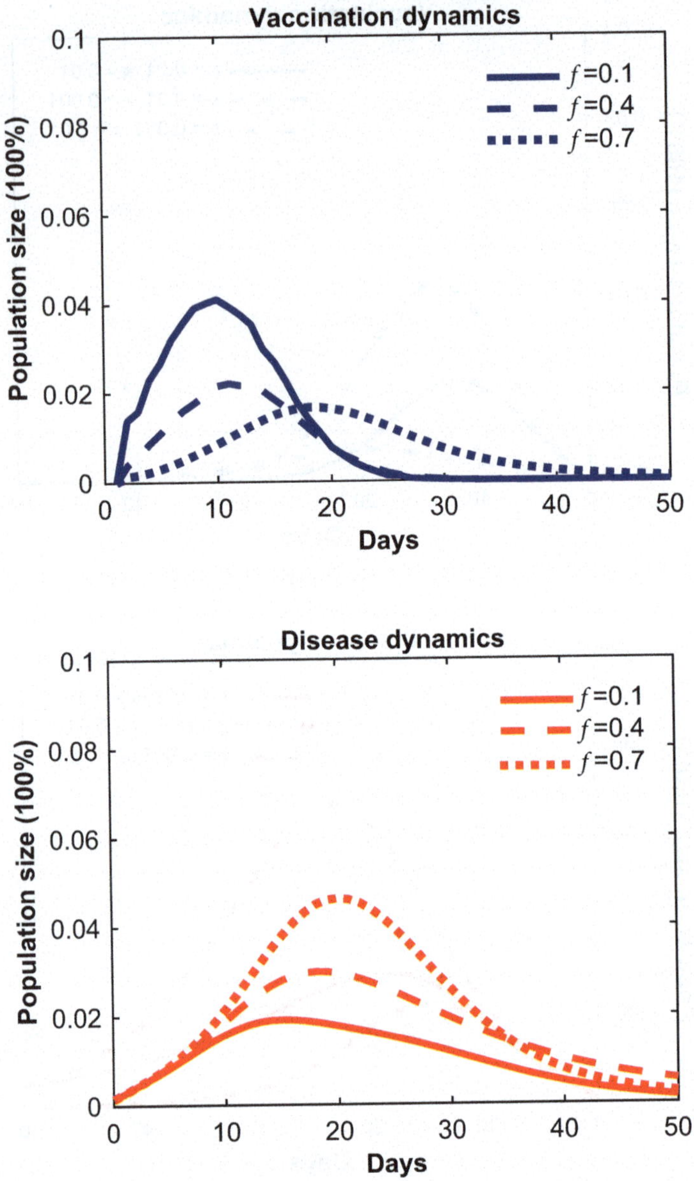

Fig. 6.7 The effect of fading coefficient (f). (**a**) The dynamics of voluntary vaccination (daily number of vaccinated individuals). (**b**) The dynamics of disease transmissions (daily number of infectious individuals)

number of vaccinated individuals peaks on day 10 at a rate of approximately 4% (Fig. 6.7a, solid curve). In contrast, the vaccination rates will be approximately 2% and peak on day 12 and day 20 if the fading coefficients are set as: $f = 0.4$ and $f = 0.7$, respectively (Fig. 6.7a, dashed curve and dotted curve). In this case, we can observe that when awareness spreads with only a weak fading effect (a smaller fading coefficient), individuals will be prompted to be vaccinated and, thus, disease transmission will be effectively prevented (Fig. 6.7b, solid curve when $f = 0.1$).

To investigate the sensitivity of our results, we consider individuals' vaccination decisions with respect to different situations of disease transmission. In Fig. 6.8, we vary the disease basic reproduction number over 1.2, 1.6, and 2.0 to represent different levels of disease infectiousness. Here, we can observe that a larger disease reproduction number will enhance individuals' vaccination rate (Fig. 6.8a, dotted curve when $R_0 = 2.0$). However, the increased levels of vaccination cannot provide adequate population-level immunity for containing disease transmission in the cases where R_0 is relatively small. The simulation results in Fig. 6.8b show that the reproduction number $R_0 = 2.0$ results in disease dynamics with the largest peak-incidence rate, i.e., nearly 6% as shown in Fig. 6.8b, dotted curve, when compared with the situation of 3% for $R_0 = 1.6$ (Fig. 6.8b, dashed curve) and 1% for $R_0 = 1.2$ (Fig. 6.8b, solid curve). These results can be explained by considering the interplay of the dynamics of disease transmission and vaccine implementation: an increased number of vaccinated individuals will trigger more reporting of the vaccine's adverse effects, which will in turn reduce individuals' vaccination willingness.

6.4 Further Remarks

It is well known that the spread of awareness about an epidemic will affect individuals' health-related behavior. For example, individuals who are aware of the risk of infection may take measures to reduce their susceptibility or distance their social contacts to protect themselves [175, 176]. In the context of vaccination, the spread of awareness about severe infections and vaccine adverse effects will affect individuals' perception of disease severity and vaccine safety and, hence, substantially change their vaccination decisions.

As a further step from the studies in the preceding chapters, here we have considered a belief-based characterization of individuals' vaccination decisions. In our model, we have correlated individuals' subjective assessment of disease severity and vaccine safety with the dynamics of disease transmission and voluntary vaccination by exploring the awareness of reported negative events. Unlike existing belief-based studies, e.g., that of Coelho et al. [177], we have characterized individuals' belief-value updates as a result of the spread of awareness in a structured host population (a Facebook-like online community). In this case, we can represent the situation in which individuals collect health-related information from online social media, and make vaccination decisions according to their obtained awareness from socially interconnected neighbors. Additionally, instead of a binary polarization of decisions,

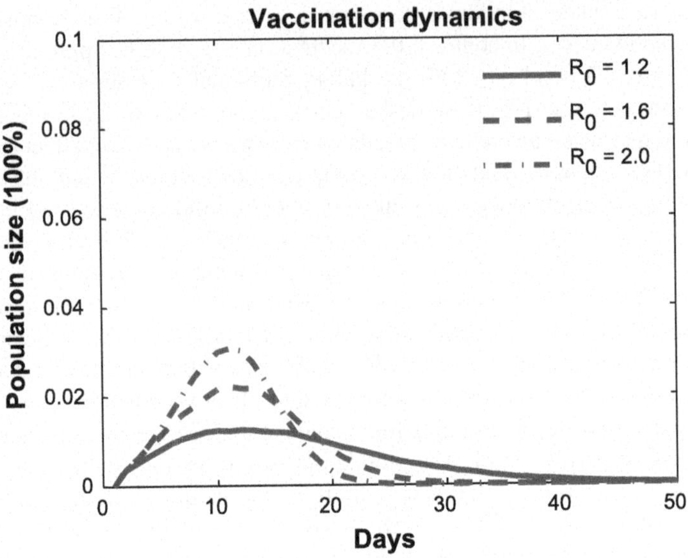

(a)

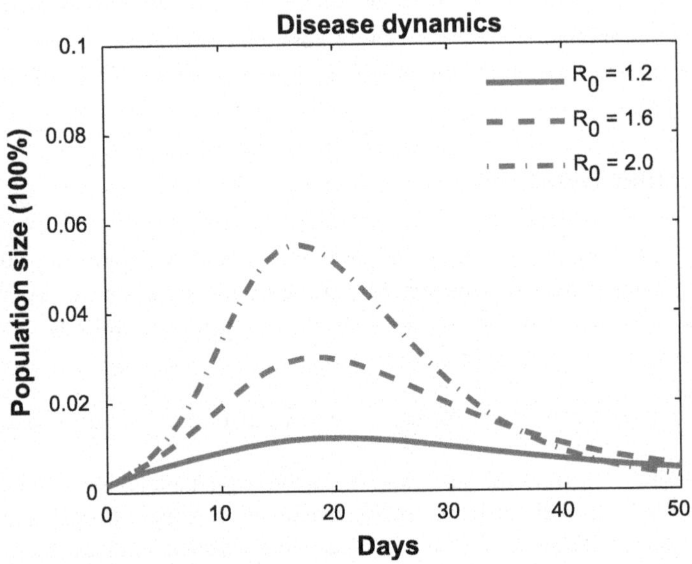

(b)

Fig. 6.8 The effect of disease reproduction number R_0. (**a**) The dynamics of voluntary vaccination (daily number of vaccinated individuals). (**b**) The dynamics of disease transmissions (daily number of infectious individuals) i.e., either to vaccinate or not, we have introduced a third decision response, "yet to decide", and associated it with the belief value of uncertainty. By doing so, we have utilized an extended DST to characterize individuals' belief-value updates in the presence of uncertainty.

By computationally characterizing the effect of the spread of awareness, this work has some practical implications for understanding individuals' vaccination decision-making and for improving the effectiveness of a vaccination program. A growing number of individuals use Internet-based communication services to obtain and share health-related information [178]. This represents the growing power of data on users' online communication to help track events in real time during an epidemic, e.g., detecting a pandemic of influenza by monitoring related queries to online search engines [179]. Salathe et al. collected individuals' Twitter updates to assess public sentiments toward a novel vaccine [180]. Henrich et al. used online comments as obtained from the websites of CBC, Vancouver Sun, and Global and Mail to capture public attitudes about the H1N1 vaccine [181]. Thus, online social media provide effective tools to access real-time data to evaluate the public perception of diseases and vaccines. These can be used to estimate public acceptance of a prospective vaccination program, which will enable public health authorities to make plans in advance to improve the effectiveness of vaccination strategies.

So far, our study has provided a modeling framework that incorporates the spread of awareness with the belief-based characterization of decision-making. It should be pointed out that the simulation results may depend on the social network used in our example (a Facebook-like online community). In our model, the spread of awareness only mediates individuals' localized interactions (i.e., between socially interconnected neighbors), while the global effect of public media is not taken into account. For future work, it will be interesting to extend the current model by adding a globalized spread of awareness, wherein each individual will become aware of a reported negative event with a certain probability. Additionally, by utilizing a DST-based characterization of vaccination decisions, the combination rule used to describe individuals' belief-value updates will determine the resulting decisions. The above-mentioned issues are worthy of further investigation.

6.5 Summary

In this chapter, to gain a more complete understanding of individuals' vaccination decision-making, we discussed how to model and evaluate the effect of the spread of vaccination and disease awareness on decisions. In doing so, we focused on a group of individuals opting for or against vaccination, in which their awareness of negative events spread via their social interactions and, thus, affected their perception of disease severity and vaccine safety. We characterized vaccination decisions in the form of individuals accepting, rejecting, or having not yet decided whether to be vaccinated, and further associated them with a set of belief values denoted Yes, No, and Uncertain, respectively. Furthermore, we extended DST to model individuals' belief updates after incorporating their awareness received from interconnected neighbors, and their corresponding changes in vaccination decisions. We studied the situation of voluntary vaccination against an influenza-like disease through a series of simulations. We examined three factors that affect individuals' vaccination decisions, namely (1) the reporting rates of negative events, (2) the coefficient of awareness fading, and (3) the disease reproduction number. By doing so, we evaluated the effect of the spread of awareness on their decisions by assessing the vaccination dynamics through the time course of disease transmission, i.e., the number of vaccinated individuals and the time point of vaccine administration.

Chapter 7
Welcome to the Era of *Systems Epidemiology*

The preceding chapters have, in many aspects, alluded to a new way of viewing and pursuing epidemiological studies in the modern context. Starting from certain empirical observations of a real-world infectious disease situation, beset by unanswered questions, we then represent it using an abstract modeling language, parameterize our models with the real-world data, and thereafter analytically and quantitatively examine the underlying associated risk factors and the implications for evidence-based policies and practice.

In this chapter, we continue to build on the above-mentioned discussions, and further generalize them through the prism of *systems epidemiology*, a term coined here to encompass the latest methodological development in epidemiology [182].

7.1 Systems Thinking in Epidemiology

We must begin by asking precisely what systems epidemiology is, and how the notion of systems is relevant to epidemiology. These key questions will be answered in the following discussions on *systems thinking*.

Systems thinking is a philosophical and methodological perspective that draws on the fundamental notions of systems theory, which views a system as an integration of components with the interacting relationships among them and the environments in which they reside [183, 184]. As mentioned by Maani et al. [185], systems thinking emphasizes two fundamental concepts, i.e., complexity and entirety. Systems complexity is generated from the structure of integrated components, i.e., how the constituent components are organized and interact with each other and with the environment. Systems entirety is derived from the dynamic behaviors of a system as a whole, which is to say, how a complex system of interacting components behaves and exhibits emergent properties at the system level, rather than being a simple behavioral aggregation of its basic components.

J. Liu, S. Xia, *Computational Epidemiology*, Health Information Science,
https://doi.org/10.1007/978-3-030-52109-7_7

Systems thinking offers a novel comprehensive perspective that examines the process of infectious disease transmission as a system, and thus having structural complexity and behavioral entirety. In such a system, the components include disease pathogens/parasites, animal/vector species, human populations, and their natural, social, and behavioral environments. The interactions among components are present, such as the ability of disease pathogens/parasites to infect and be transmitted between and within animal/vector species and human populations. The interacting relationships of components with their environments can be described as those components' responses to potential environmental changes, for example, biomedical genetic mutations of pathogens or parasites as a result of drug resistance selection, animal/vector population fluctuations due to climate changes, and behavioral changes affecting people's exposure due to their socio-economic conditions. The emergent behaviors of such a system, i.e., the emerging and re-emerging infectious diseases, depend on the integrated effects of all of the constituting components, including evolving microbial pathogens, zoonotic vector exposures, environmental changes, and human behaviors.

Based on the above-mentioned perspective of systems thinking, the study of infectious diseases can go beyond the conventional methods that are usually confined by their disciplinary boundaries, such as the statistical analysis of disease occurrences or laboratory research on pathogens. Intervention measures to combat infectious diseases will thus be designed to modify the emergent behaviors at the system level, by exploiting interdisciplinary methods to address systems complexity. Toward this end, we need a set of novel modeling and analytical tools drawing on the concepts of systems thinking. This will enable epidemiologists to develop and deploy more effective intervention measures.

7.2 Systems Modeling in Principle

The complex systems approach is a holistic approach intended to model, characterize, explain, and predict the emergent behaviors of a system with reference to its constituting complexity, which is difficult to derive or compute using conventional top-down reductionist approaches [186, 187]. As stated by Liu [188], such an approach pays special attention to the following three objectives:

- **Systems Modeling**
  The systems-modeling step provides a blueprint/framework that is abstracted and replicated from real-world observations in the languages of mathematical/computational characterizations. To model a system in this way requires identification, abstraction, and reproduction of certain observations, which is the starting point for the following steps of systems exploration and problem solving. In systems modeling, the basic components of a model, also known as entities, are the basic constituents of a complex system, which directly or indirectly interact among themselves and with their environments, based on certain predefined

or known mechanisms or principles. For example, in a network-based disease model, nodes represent individual humans whereas links represent the routes of disease transmission. Diseases can be transmitted from one node to another due to the predefined contact interactions [189]. Interrelationships exist among entities and their local and global environments, through which the complex system, as a whole, exhibits structural and behavioral complexity at and across various scales. Emergence, which is the dynamic transmission of infectious diseases, is defined in terms of the system-level patterns and regularities arising from the dynamics of a group of interacting entities, as generated from the reciprocally coupled and dynamically changing interrelations of the entities at multiple scales.

- **Systems Exploration**
  Systems exploration presents a set of analytical tools to understand the operating mechanisms underlying a complex system and, furthermore, find explanations and make predictions about the observed systems' dynamics. To uncover the operating mechanisms behind the observations, we perform systems modeling to characterize or simulate the real system. For example, we can use the SIR model to characterize the dynamics of disease transmission in a human population. Then, by comparing the real-world observations and the synthetic simulation, the models and interaction mechanisms of the system can be fine-tuned, as reflected in adjustments of the relevant model parameters, and the structure and behavior of the interacting entities. For example, when the incubation period is taken into consideration, the SIR model can be modified into the susceptible-exposed-infectious-recovered (SEIR) model with an additional latency compartment of E (exposed).
- **Problem Solving**
  Problem solving emphasizes the ability of the complex systems approach to independently find ways to achieve adaptive solutions that are well suited to the current problem settings. The ultimate goal is to develop a set of analytical algorithms that can adjust their own parameters for different application domains. For example, adaptive evolutionary algorithms can be used to automatically tune parameters related to the developed systems being modeled or the proposed operating mechanisms. Constrained optimization algorithms are dedicated to finding the optimal solutions for resource allocation.

The complex systems approach can be used not only to build a modeling framework for mapping real-world observations/phenomena in analytical languages (i.e., mathematical and computational models), but also to reveal the operating mechanisms behind a complex system. Problem solving is the advance application of systems modeling and exploration, with respect to the specific domain problem set.

7.3 Systems Modeling in Practice

In combating infectious diseases, the complex systems approach can help us understand how the systems of infectious diseases are organized, in terms of the causal relationships and the factors affecting disease prevalence, and how such systems behave in time and space, by revealing the spatio-temporal distributions of disease occurrences, and how the diseases can be better mitigated and eradicated by developing more effective solutions for infectious disease control.

Figure 7.1 provides a schematic framework outlining the four essential steps for applying the complex systems approach to epidemiological studies of infectious diseases.

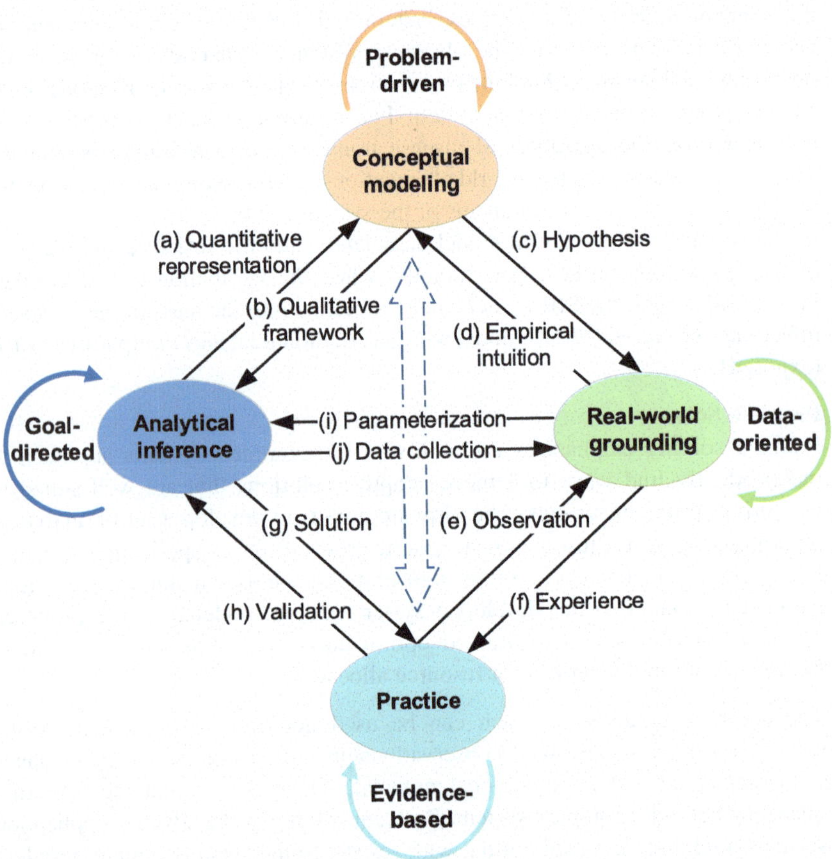

Fig. 7.1 The four essential steps (in ovals) for performing the complex systems approach in combating infectious diseases. The directional arrows show their functional interrelationships

- **Problem-Driven Conceptual Modeling** first translates the real-world problems in an epidemiological domain into conceptual models in a theoretical or computational domain, which aim to describe the systems components in infectious disease transmission, their impact factors, and interaction relationships.
- **Data-Oriented Real-World Grounding** then concentrates on discovering ways to embody conceptual models, through model parameterization, by obtaining and utilizing real-world data and statistically analyzing the real-world observations.
- **Goal-Directed Analytical Inference** is devoted to further developing analytical methods and solutions addressing specific real-world tasks of disease surveillance and control; that is, to finding appropriate methods and solutions to meet specific goals.
- **Evidence-Based Practice** proceeds to the implementation, validation, and improvement of the developed analytical solutions, aiming to bridge the theoretical or computational analysis and the real world.

Specifically, the goal of the conceptual modeling step is to build theoretical or computational prototypes of infectious disease systems, which can be used to represent the real-world problems. For example, networks, consisting of nodes and links between nodes, have been utilized to characterize the dynamics of infectious diseases with various diffusion paths [190–192].

Based on the existing understanding and theoretical/empirical knowledge about infectious diseases and the related impact factors, mathematical and computational models can be used as a conceptual framework to reproduce disease dynamics. For example, the compartmental models are well suited to characterizing disease dynamics in several host populations, such as in the case of influenza [193]. Meanwhile, the network models or metapopulation-based models are more suitable for representing disease diffusion due to human movement, such as the importation of malaria cases in remote or cross-border areas [71, 194, 195]. To achieve the above, we need to perform model selection with reference to the specific characteristics of the current epidemiological problems.

As can be seen, conceptual modeling depends on simplifications and abstractions about the operating mechanisms of infectious diseases, which also necessitates the use of hypotheses for data collection in the real-world grounding step (function c in Fig. 7.1). This step also provides a theoretical or computational qualitative framework for performing analytical inference methods (function b). The goal of real-world grounding is to collect and analyze data from multiple sources, aiming at a more comprehensive, multidisciplinary understanding of the structural interrelationships and behavioral mechanisms of real-world infectious disease systems. For example, in the case of influenza, the demographic profiles and contact structures of a human host population can be used to model disease transmission among different human groups [196, 197]. In the case of malaria, environmental factors, such as rainfall and temperature, can be identified from various sources, and incorporated into a causal model for examining the effect of disease vector population over time [198].

The products of this data-oriented real-world grounding can in turn provide empirical intuitions for conceptual modeling (function d), generate experience-based rules or principles to guide the practical implementation of infectious disease-control measures (function f), and parameterize variables in performing inference algorithms (function i). The real-world grounding step involves multi-disciplinary data-fusion and knowledge discovery from massively accumulated data. For example, based on the reported infection cases from the 62 towns in Yunnan province of China, the underlying transmission networks of P. vivax can be inferred for public health policy makers to accurately predict the geographic patterns of malaria spread [199]. In order to evaluate the potential risks of H7N9 infection, data about bird migration and poultry distribution can be collected and utilized [200].

Based on the developed models and collected data, the analytical inference step provides a series of specific problem-solving methods and solutions, which can be used as analytical tools to address the real-world problems that are taken into account in the conceptual modeling step. The gaps between the desired situations (goals) and the current situations (status quo) in disease active surveillance and control necessitate the use of inference methods to find an improved solution. Analytical inferences will provide a set of quantitative representations for conceptual modeling (function a). Furthermore, the end products of this step can also inform solutions for the practical realization of infectious disease control (function g) and guide the data collection in the real-world grounding step (function j). For example, network inference methods can be used to identify the spatio-temporal patterns of dengue transmission [201] as well as to reveal the key bird species and geographic hotspots of avian influenza A (H7N9) [202]. Poisson regression methods can be developed for epidemic prediction by integrating both inter-regional and external, environmental, and social impact factors [203]. The methods of group sparse Bayesian learning and partially observable reinforcement learning can be utilized to achieve active surveillance of infectious diseases [204, 205].

The fourth step of evidence-based practice concerns the application and validation of the developed solutions in the real-world practice of infectious disease surveillance and control. The goal of this step is twofold: (1) guiding the practice of disease control and prevention (function e); and (2) validating and improving the applied analytical methods (function h). For example, intervention and surveillance planning can help public health authorities to know how to distribute their very limited resources to high-priority regions so as to maximize the outcomes of active surveillance [55, 206]. Risk ranking can also help identify the relative risks of malaria among various villages in remote or cross-border areas [198, 207]. At the same time, as more data are accumulated, the results of risk ranking will become more precise and reliable [208].

The feedback from field studies will help validate the analytical results and determine if the selected models and adopted inference methods truly represent the real-world scenario and, thus, address the real-world problems. In other words, theoretical analysis and results will be used to guide the practice of infectious disease control, which will in turn validate or improve the developed models and inference methods [209].

7.4 Toward Systems Epidemiology

Systems epidemiology will be an interdisciplinary effort that will help provide a deeper understanding of diseases and mitigation strategies. It will enable us to pursue comprehensive epidemiological inquiries, throughout the evolutionary life-cycle of *problem-driven conceptual modeling*, *data-oriented real-world grounding*, *goal-directed analytical inference*, and *evidence-based practice*, as detailed above. Each step will offer either a sound validation of, or a new insight into, another step, thus bringing our understanding of the study problem closer to reality. The emerging computational modeling and analytics tools and artificial intelligence technologies will further empower such an endeavor.

Let us make this paradigm shift now—Welcome to the new era!

References

1. Last JM, Harris SS, Thuriaux MC, et al. A Dictionary of Epidemiology. New York: Oxford University Press; 2014.
2. MacMahon B, Trichopoulos D. Epidemiology: Principles and Methods. Philadelphia: Lippincott Williams and Wilkins; 1997.
3. Merrill RM. Introduction to Epidemiology. Burlington: Jones and Bartlett Learning; 2015.
4. Rothman KJ. Lessons from John Graunt. Lancet. 1996;347(8993):37–9. https://doi.org/10.1016/S0140-6736(96)91562-7.
5. Koutouvidis N, Marketos S, Beveridge A. The contribution of Thomas Sydenham (1624-1689) to the evolution of psychiatry. Hist Psychiatry. 1995;6(24):513–20. https://doi.org/10.1177/0957154X9500602408.
6. Cameron D, Jones IG. John Snow, the broad street pump and modern epidemiology. Int J Epidemiol. 1983;12(4):393–96. https://doi.org/10.1093/ije/12.4.393.
7. Smith DL, Battle KE, Hay SI, et al. Ross, Macdonald, and a theory for the dynamics and control of mosquito-transmitted pathogens. PLoS Pathog. 2012;8(4):e1002588. https://doi.org/10.1371/journal.ppat.1002588.
8. Zadoks J. Methodology of epidemiological research. Ann Rev Phytopathol. 1972;10(1):253–76. https://doi.org/10.1146/annurev.py.10.090172.001345.
9. Rothman KJ, Greenland S, Lash TL. Modern Epidemiology. Philadelphia: Lippincott Williams & Wilkins; 2008.
10. Miettinen OS. Theoretical Epidemiology: Principles of Occurrence Research in Medicine. Wiley: New York; 1985
11. Marathe M, Vullikanti AKS. Computational epidemiology. Commun ACM. 2013;56(7):88-96. https://doi.org/10.1145/2483852.2483871.
12. Morse SS, Mazet JA, Woolhouse M, et al. Prediction and prevention of the next pandemic zoonosis. Lancet. 2012;380(9857):1956–65. https://doi.org/10.1016/s0140-6736(12)61684-5.
13. Morens DM, Folkers GK, Fauci AS. The challenge of emerging and re-emerging infectious diseases. Nature. 2004;430(6996):242. https://doi.org/10.1038/nature02759.
14. Fauci AS, Morens DM. The perpetual challenge of infectious diseases. N Engl J Med. 2012;366(5):454–61. https://doi.org/10.1056/NEJMc1204960SA2.
15. Protopopoff N, Van Bortel W, Speybroeck N, et al. Ranking malaria risk factors to guide malaria control efforts in African highlands. PLoS One. 2009;4(11):e8022. https://doi.org/10.1371/journal.pone.0008022.
16. Flahault A, Zylberman P. Influenza pandemics: Past, present and future challenges. Public Health Rev. 2010;32(1):319–40. https://doi.org/10.1007/BF03391605.

J. Liu, S. Xia, *Computational Epidemiology*, Health Information Science,
https://doi.org/10.1007/978-3-030-52109-7

17. Pearce N, Merletti F. Complexity, simplicity, and epidemiology. Int J Epidemiol. 2006;35(3):515–9. https://doi.org/10.1093/ije/dyi322.
18. Wu JT, Leung K, Leung GM. Nowcasting and forecasting the potential domestic and international spread of the 2019-nCoV outbreak originating in Wuhan, China: A modelling study. Lancet. 2020;395(10225):689–97. https://doi.org/10.1016/S0140-6736(20)30260-9.
19. Ferguson HM, Dornhaus A, Beeche A, et al. Ecology: A prerequisite for malaria elimination and eradication. PLoS Med. 2010;7(8):e1000303. https://doi.org/10.1371/journal.pmed.1000303.
20. Martens P, Hall L. Malaria on the move: Human population movement and malaria transmission. Emerg Infect Dis. 2000;6(2):103. https://doi.org/10.3201/eid0602.000202.
21. Mooney SJ, Westreich DJ, El-Sayed AM. Epidemiology in the era of big data. Epidemiology. 2015;26(3):390. https://doi.org/10.1097/EDE.0000000000000274.
22. Hay SI, George DB, Moyes CL, et al. Big data opportunities for global infectious disease surveillance. PLoS Med. 2013;10(4):e1001413. https://doi.org/10.1371/journal.pmed.1001413.
23. Lobitz B, Beck L, Huq A, et al. Climate and infectious disease: Use of remote sensing for detection of Vibrio cholerae by indirect measurement. Proc Natl Acad Sci. 2000;97(4):1438–43. https://doi.org/10.1073/pnas.97.4.1438.
24. Brooker S, Michael E. The potential of geographical information systems and remote sensing in the epidemiology and control of human helminth infections. Adv Parasitol. 2000;47:245–88. https://doi.org/10.1016/S0065-308X(00)47011-9.
25. Cook S, Conrad C, Fowlkes AL, et al. Assessing Google flu trends performance in the United States during the 2009 influenza virus A (H1N1) pandemic. PLoS One. 2011;6(8):e23610. https://doi.org/10.1371/journal.pone.0023610.
26. Althouse BM, Ng YY, Cummings DA. Prediction of dengue incidence using search query surveillance. PLoS Negl Trop Dis. 2011;5(8):e1258. https://doi.org/10.1371/journal.pntd.0001258.
27. Longini IM, Nizam A, Xu S, et al. Containing pandemic influenza at the source. Science. 2005;309(5737):1083–7. https://doi.org/10.1126/science.1115717.
28. Ferguson NM, Cummings DA, Cauchemez S, et al. Strategies for containing an emerging influenza pandemic in Southeast Asia. Nature. 2005;437(7056):209–14. https://doi.org/10.1038/nature04017.
29. Ferguson NM, Cummings DA, Fraser C, et al. Strategies for mitigating an influenza pandemic. Nature. 2006;442(7101):448–52. https://doi.org/10.1038/nature04795.
30. Cauchemez S, Valleron AJ, Boelle PY, et al. Estimating the impact of school closure on influenza transmission from Sentinel data. Nature. 2008;452(7188):750–4. https://doi.org/10.1038/nature06732.
31. Wu JT, Cowling BJ, Lau EHY, et al. School closure and mitigation of pandemic (H1N1) 2009, Hong Kong. Emerg Infect Dis. 2010;16(3):538–41. https://doi.org/10.3201/eid1603.091216.
32. Fine PE. Herd immunity: History, theory, practice. Epidemiol Rev. 1993;15(2):265–302. https://doi.org/10.1093/oxfordjournals.epirev.a036121.
33. Zimmer SM, Burke DS. Historical perspective - emergence of influenza A (H1N1) viruses. N Engl J Med. 2009;361(3):279–85. https://doi.org/10.1056/NEJMra0904322.
34. Hay AJ, Gregory V, Douglas AR, et al. The evolution of human influenza viruses. Philos Trans R Soc Lond B Biol Sci. 2001;356(1416):1861–70. https://doi.org/10.1098/rstb.2001.0999.
35. Partridge J, Kieny MP, World Health Organization H1N1 Influenza Vaccine Task Force. Global production of seasonal and pandemic (H1N1) influenza vaccines in 2009-2010 and comparison with previous estimates and global action plan targets. Vaccine. 2010;28(30):4709–12. https://doi.org/10.1016/j.vaccine.2010.04.083.
36. Oshitani H, Kamigaki T, Suzuki A. Major issues and challenges of influenza pandemic preparedness in developing countries. Emerg Infect Dis. 2008;14(6):875. https://doi.org/10.3201/eid1406.070839.

37. World Health Organization. WHO guidelines on the use of vaccines and antivirals during influenza pandemics; 2004. Accessed 10 Aug 2004. Website, Available from: http://www.who.int/csr/resources/publications/influenza/WHO_CDS_CSR_RMD_2004_8/en/index.html.
38. Gangarosa EJ, Galazka A, Wolfe C, et al. Impact of anti-vaccine movements on pertussis control: The untold story. Lancet. 1998;351(9099):356–61. https://doi.org/10.1109/4.772418.
39. Roberts RJ, Sandifer QD, Evans MR, et al. Reasons for non-uptake of measles, mumps, and rubella catch up immunisation in a measles epidemic and side effects of the vaccine. Br Med J. 1995;310(6995):1629–39. https://doi.org/10.1136/BMJ.310.6995.1629.
40. Anderson P. Another media scare about MMR vaccine hits Britain. Br Med J. 1999;318(7198):1578. https://doi.org/10.1136/BMJ.318.7198.1578.
41. Poland GA, Jacobson RM. Understanding those who do not understand: A brief review of the anti-vaccine movement. Vaccine. 2001;19(17-19):2440–5. https://doi.org/10.1016/S0264-410X(00)00469-2.
42. Hobson-West P. 'Trusting blindly can be the biggest risk of all': Organised resistance to childhood vaccination in the UK. Sociol Health Illn. 2007;29(2):198–215. https://doi.org/10.1111/j.1467-9566.2007.00544x.
43. Jansen VA, Stollenwerk N, Jensen HJ, et al. Measles outbreaks in a population with declining vaccine uptake. Science. 2003;301(5634):804. https://doi.org/10.1126/science.1086726.
44. Poland GA, Jacobson RM. The age-old struggle against the antivaccinationists. N Engl J Med. 2011;364(2):97–9. https://doi.org/10.1056/NEJMp1010594.
45. Streefland PH. Public doubts about vaccination safety and resistance against vaccination. Health Policy. 2001;55(3):159–72. https://doi.org/10.1016/S0168-8510(00)00132-9.
46. Freed GL, Clark SJ, Hibbs BF, et al. Parental vaccine safety concerns: The experiences of pediatricians and family physicians. Am J Prev Med. 2004;26(1):11–4. https://doi.org/10.1016/j.amepre.2003.09.004.
47. Bauch CT, Galvani AP, Earn DJD. Group interest versus self-interest in smallpox vaccination policy. Proc Natl Acad Sci. 2003;100(18):10564–7. https://doi.org/10.1073/pnas.1731324100.
48. Galvani AP, Reluga TC, Chapman GB. Long-standing influenza vaccination policy is in accord with individual self-interest but not with the utilitarian optimum. Proc Natl Acad Sci. 2007;104(13):5692–7. https://doi.org/10.1073/pnas.0606774104.
49. Jadhav SS, Gautam M, Gairola S. Emerging markets and emerging needs: Developing countries vaccine manufacturers' perspective and its current status. Biologicals. 2009;37(3):165–8. https://doi.org/10.1016/j.biologicals.2009.02.009.
50. Friede M, Palkonyay L, Alfonso C, et al. WHO initiative to increase global and equitable access to influenza vaccine in the event of a pandemic: Supporting developing country production capacity through technology transfer. Vaccine. 2011;29(S1):A2–7. https://doi.org/10.1016/j.vaccine.2011.02.079.
51. Keelan J, Pavri V, Balakrishnan R, et al. An analysis of the human papilloma virus vaccine debate on MySpace blogs. Vaccine. 2010;28(6):1535–40. https://doi.org/10.1016/j.vaccine.2009.11.060.
52. Liu Y, Gu Z, Xia S, et al. What are the underlying transmission patterns of COVID-19 Outbreak? An age-specific social contact characterization. EClinicalMedicine. 2020;100354. https://doi.org/10.1016/j.eclinm.2020.100354.
53. Xia S, Liu J. A computational approach to characterizing the impact of social influence on individuals' vaccination decision making. PLoS One. 2013;8(4):e60373. https://doi.org/10.1371/journal.pone.0060373.
54. Xia S, Liu J. A belief-based model for characterizing the spread of awareness and its impacts on individuals' vaccination decisions. J R Soc Interface. 2014;11(94):20140013. https://doi.org/10.1098/rsif.2014.0013.
55. Xia S, Liu J, Cheung W. Identifying the relative priorities of subpopulations for containing infectious disease spread. PLoS One. 2013;8(6):e65271. https://doi.org/10.1371/journal.pone.0065271.

56. Hethcote HW. The mathematics of infectious diseases. SIAM Rev. 2000;42(4):599–653. https://doi.org/10.1137/S0036144500371907.
57. Keeling MJ, Rohani P. Modeling Infectious Diseases in Humans and Animals. Princeton: Princeton University Press; 2011.
58. Edmunds WJ, O'callaghan C, Nokes D. Who mixes with whom? A method to determine the contact patterns of adults that may lead to the spread of airborne infections. Proc R Soc B Biol Sci. 1997;264(1384):949–57. https://doi.org/10.1098/rspb.1997.0131.
59. Diekmann O, Heesterbeek JAP, Metz JA. On the definition and the computation of the basic reproduction ratio R_0 in models for infectious diseases in heterogeneous populations. J Math Biol. 1990;28(4):365–82. https://doi.org/10.1007/BF00178324.
60. Heesterbeek JAP. A brief history of R_0 and a recipe for its calculation. Math Philos Found Biol Biomed Science. 2002;50(3):189–204. https://doi.org/10.1023/A:1016599411804.
61. Bailey NT. The Mathematical Theory of Infectious Diseases and Its Applications. London: Griffin; 1975.
62. Anderson RM, May RM. Infectious Diseases of Humans: Dynamics and Control. New York: Oxford University Press; 1992.
63. Coburn BJ, Wagner BG, Blower S. Modeling influenza epidemics and pandemics: Insights into the future of swine flu (H1N1). BMC Med. 2009;7(1):30. https://doi.org/10.1186/1741-7015-7-30.
64. Keeling MJ, Grenfell BT. Understanding the persistence of measles: Reconciling theory, simulation and observation. Proc R Soc B Biol Sci. 2002;269(1489):335–43. https://doi.org/10.1098/rspb.2001.1898.
65. Keeling MJ, Eames KT. Networks and epidemic models. J R Soc Interface. 2005;2(4):295–307. https://doi.org/10.1098/rsif.2005.0051.
66. Keeling M. The implications of network structure for epidemic dynamics. Theor Popul Biol. 2005;67(1):1–8. https://doi.org/10.1016/j.tpb.2004.08.002.
67. Tan Q, Liu Y, Liu J. Motif-aware diffusion network inference. Int J Data Sci Anal. 2018;1–13. https://doi.org/10.1007/978-3-319-93040-4_50.
68. Dangerfield CE, Ross JV, Keeling MJ. Integrating stochasticity and network structure into an epidemic model. J R Soc Interface. 2008;6(38):761–774. https://doi.org/10.1098/rsif.2008.0410.
69. Salathe M, Kazandjieva M, Lee JW, et al. A high-resolution human contact network for infectious disease transmission. Proc Natl Acad Sci. 2010;107(51):22020–5. https://doi.org/10.1073/pnas.1009094108.
70. Grenfell B, Harwood J. (Meta) population dynamics of infectious diseases. Trends Ecol Evol. 1997;12(10):395–9. https://doi.org/10.1088/0034-4885/77/2/026602.
71. Yang X, Liu J, Cheung WKW, et al. Inferring metapopulation based disease transmission networks. In: 18th Pacific-Asia Conference on Knowledge Discovery and Data Mining. Berlin: Springer; 2014. p. 385–99.
72. Lloyd-Smith JO, Schreiber SJ, Kopp PE, et al. Superspreading and the effect of individual variation on disease emergence. Nature. 2005;438(7066):355–9. https://doi.org/10.1038/nature0415.
73. Dushoff J, Plotkin JB, Viboud C, et al. Vaccinating to protect a vulnerable subpopulation. PLoS Med. 2007;4(5):e174. https://doi.org/10.1371/journal.pmed.0040174.
74. Wallinga J, Edmunds WJ, Kretzschmar M. Perspective: Human contact patterns and the spread of airborne infectious diseases. Trends Microbiol. 1999;7(9):372–7. https://doi.org/10.1016/S0966-842X(99)01546-2.
75. Stehle J, Voirin N, Barrat A, et al. High-resolution measurements of face-to-face contact patterns in a primary school. PLoS One. 2011;6(8):e23176. https://doi.org/10.1371/journal.pone.0023176.
76. Haydon DT, Chase-Topping M, Shaw D, et al. The construction and analysis of epidemic trees with reference to the 2001 UK foot-and-mouth outbreak. Proc R Soc B Biol Sci. 2003;270(1511):121–7. https://doi.org/10.1098/rspb.2002.2191.

77. Riley S, Fraser C, Donnelly CA, et al. Transmission dynamics of the etiological agent of SARS in Hong Kong: Impact of public health interventions. Science. 2003;300(5627):1961–6. https://doi.org/10.1126/science.1086478.
78. Eames KT, Keeling MJ. Contact tracing and disease control. Proc R Soc B Biol Sci. 2003;270(1533):2565–71. https://doi.org/10.1098/rspb.2003.2554.
79. Bearman PS, Moody J, Stovel K. Chains of affection: The structure of adolescent romantic and sexual networks. Sci Eng Ethics. 2004;110(1):44–91.
80. De P, Singh AE, Wong T, et al. Sexual network analysis of a gonorrhoea outbreak. Sex Transm Infect. 2004;80(4):280–5. https://doi.org/10.1136/sti.2003.007187.
81. Lalvani A, Pathan AA, Durkan H, et al. Enhanced contact tracing and spatial tracking of Mycobacterium tuberculosis infection by enumeration of antigen-specific T cells. Lancet. 2001;357(9273):2017–21. https://doi.org/10.1016/S0140-6736(00)05115-1.
82. Eichner M. Case isolation and contact tracing can prevent the spread of smallpox. Am J Epidemiol. 2003;158(2):118–128. https://doi.org/10.1093/aje/kwg104.
83. Mossong J, Hens N, Jit M, et al. Social contacts and mixing patterns relevant to the spread of infectious diseases. PLoS Med. 2008;5(3):e74. https://doi.org/10.1371/journal.pmed.0050074.
84. Read JM, Eames KT, Edmunds WJ. Dynamic social networks and the implications for the spread of infectious disease. J R Soc Interface. 2008;5(26):1001–7. https://doi.org/10.1098/rsif.2008.0013.
85. Hens N, Goeyvaerts N, Aerts M, et al. Mining social mixing patterns for infectious disease models based on a two-day population survey in Belgium. BMC Infect Dis. 2009;9(1):5. https://doi.org/10.1186/1471-2334-9-5.
86. Doherty IA, Padian NS, Marlow C, et al. Determinants and consequences of sexual networks as they affect the spread of sexually transmitted infections. J Infect Dis. 2005;191 Suppl 1:S42–54. https://doi.org/10.1086/425277.
87. Liljeros F, Edling CR, Amaral LAN. Sexual networks: Implications for the transmission of sexually transmitted infections. Microbes Infect. 2003;5(2):189–96. https://doi.org/10.1016/s1286-4579(02)00058-8.
88. Potterat J, Phillips-Plummer L, Muth S, et al. Risk network structure in the early epidemic phase of HIV transmission in Colorado Springs. Sex Transm Dis. 2002;78 Suppl 1:i159–63. https://doi.org/10.1136/sti.78.suppl1.i159.
89. Yu Z, Liu J, Zhu X. Inferring a district-based hierarchical structure of social contacts from census data. PLoS One. 2015;10(2):e0118085. https://doi.org/10.1371/journal.pone.0118085.
90. Halloran ME, Longini IM, Nizam A, et al. Containing bioterrorist smallpox. Science. 2002;298(5597):1428–32. https://doi.org/10.1126/science.1074674.
91. Meyers LA, Pourbohloul B, Newman ME, et al. Network theory and SARS: Predicting outbreak diversity. J Theor Biol. 2005;232(1):71–81. https://doi.org/10.1016/j.jtbi.2004.07.026.
92. Eubank S, Guclu H, Kumar VA, et al. Modelling disease outbreaks in realistic urban social networks. Nature. 2004;429(6988):180–4. https://doi.org/doi:10.1038/nature02541.
93. Eubank S, Kumar VA, Marathe MV, et al. Structure of social contact networks and their impact on epidemics. DIMACS Ser Discrete Math Theor Comput Sci. 2006;70:181–214. https://doi.org/10.1090/dimacs/070/09.
94. The Hong Kong Government. 1st H1N1 case confirmed in HK; 2009. Accessed 1 May 2009. Website, Available from: https://www.news.gov.hk/isd/ebulletin/en/category/healthandcommunity/090501/html/090501en05004.htm.
95. Center for Health Protection. Summary report on the surveillance of adverse events following HSI immunisation and expert group's comment on the safety of HSI vaccine in Hong Kong; 2010. Accessed 1 April 2011. Website, Available from: http://www.chp.gov.hk/files/pdf/hsi_vaccine_aefi_report_en.pdf.
96. Center for Health Protection. Swine and seasonal flu monitor; 2010. Accessed 1 Oct 2010. Website, Available from: https://www.chp.gov.hk/files/pdf/ssfm_26_11_09.pdf.

97. Census and Statistics Department of the Hong Kong Government. 2011 Hong Kong population census; 2011. Accessed 1 April 2020. Website, Available from: http://www.census2011.gov.hk.
98. Kwok KO, Leung GM, Riley S. Modelling the proportion of influenza infections within households during pandemic and non-pandemic years. PLoS One. 2011;6(7):e22089. https://doi.org/10.1371/journal.pone.0022089.
99. Fumanelli L, Ajelli M, Manfredi P, et al. Inferring the structure of social contacts from demographic data in the analysis of infectious diseases spread. PLoS Comput Biol. 2012;8(9):e1002673. https://doi.org/10.1371/journal.pcbi.1002673.
100. Merler S, Ajelli M, Pugliese A, et al. Determinants of the spatiotemporal dynamics of the 2009 H1N1 pandemic in Europe: Implications for real-time modelling. PLoS Comput Biol. 2011;7(9):e1002205. https://doi.org/10.1371/journal.pcbi.1002205.
101. Wu JT, Ma ES, Lee CK, et al. The infection attack rate and severity of 2009 pandemic H1N1 influenza in Hong Kong. Clin Infect Dis. 2010;51(10):1184–91. https://doi.org/10.1086/656740.
102. Wu JT, Ho A, Ma ESK, et al. Estimating infection attack rates and severity in real time during an influenza pandemic: Analysis of serial cross-sectional serologic surveillance data. PLoS Med. 2011;8(10):e1001103. https://doi.org/10.1371/journal.pmed.1001103.
103. Cowling BJ, Lau MSY, Ho LM, et al. The effective reproduction number of pandemic influenza: Prospective estimation. Epidemiology. 2010;21(6):842–6. https://doi.org/10.1097/EDE.0b013e3181f20977.
104. Cowling BJ, Fang VJ, Riley S, et al. Estimation of the serial interval of influenza. Epidemiology. 2009;20(3):344–7. https://doi.org/10.1097/EDE.0b013e31819d1092.
105. Cauchemez S, Ferguson NM, Wachtel C, et al. Closure of schools during an influenza pandemic. Lancet Infect Dis. 2009;9(8):473–81. https://doi.org/10.1016/S1473-3099(09)70176-8.
106. Wallinga J, van Boven M, Lipsitch M. Optimizing infectious disease interventions during an emerging epidemic. Proc Natl Acad Sci. 2010;107(2):923–8. https://doi.org/10.1073/pnas.0908491107.
107. Liu J, Xia S. Toward effective vaccine deployment: A systematic study. J Med Syst. 2011;35(5):1153–64. https://doi.org/10.1007/s10916-011-9734-x.
108. Keeling MJ, White PJ. Targeting vaccination against novel infections: Risk, age and spatial structure for pandemic influenza in Great Britain. J R Soc Interface. 2011;8(58):661–70. https://doi.org/10.1098/rsif.2010.0474.
109. Hashemian M, Qian W, Stanley KG, et al. Temporal aggregation impacts on epidemiological simulations employing microcontact data. BMC Med Inform Decis Mak. 2012;12(1):132. https://doi.org/10.1186/1472-6947-12-132.
110. Ferguson N. Capturing human behaviour. Nature. 2007;446(7137):733. https://doi.org/10.1038/446733a.
111. John TJ, Samuel R. Herd immunity and herd effect: New insights and definitions. Eur J Epidemiol. 2000;16(7):601–6. https://doi.org/10.1023/A:1007626510002.
112. Mounier-Jack S, Coker RJ. How prepared is Europe for pandemic influenza? Analysis of national plans. Lancet. 2006;367(9520):1405–11. https://doi.org/10.1016/S0140-6736(06)68511-5.
113. Straetemans M, Buchholz U, Reiter S, et al. Prioritization strategies for pandemic influenza vaccine in 27 countries of the European Union and the Global Health Security Action Group: A review. BMC Public Health. 2007;7(1):236. https://doi.org/10.1186/1471-2458-7-236.
114. Ng S, Wu P, Nishiura H, et al. An analysis of national target groups for monovalent 2009 pandemic influenza vaccine and trivalent seasonal influenza vaccines in 2009-10 and 2010-11. BMC Infect Dis. 2011;11(1):230. https://doi.org/10.1186/1471-2334-11-230.
115. Coker R, Mounier-Jack S. Pandemic influenza preparedness in the Asia Pacific region. Lancet. 2006;368(9538):886–9. https://doi.org/10.1016/S0140-6736(06)69209-X.
116. Germann TC, Kadau K, Longini IM, et al. Mitigation strategies for pandemic influenza in the United States. Proc Natl Acad Sci. 2006;103(15):5935–40. https://doi.org/10.1073/pnas.0601266103.

117. Medlock J, Galvani AP. Optimizing in influenza vaccine distribution. Science. 2009;325(5948):1705–8. https://doi.org/10.1126/science.1175570.
118. Tuite AR, Fisman DN, Kwong JC, et al. Optimal pandemic influenza vaccine allocation strategies for the Canadian population. PLoS One. 2010;5(5):e10520. https://doi.org/10.1371/journal.pone.0010520.
119. Yang Y, Sugimoto JD, Halloran E, et al. The transmissibility and control of pandemic influenza A (H1N1) virus. Science. 2009;326(5953):729–33. https://doi.org/10.1126/science.1177373.
120. Keeling MJ, White PJ. Targeting vaccination against novel infections: Risk, age and spatial structure for pandemic influenza in Great Britain. J R Soc Interface. 2011;8(58):661–70. https://doi.org/10.1098/rsif.2010.0474.
121. Medlock J, Meyers LA, Galvani A. Optimizing allocation for a delayed influenza vaccination campaign. PLoS Curr. 2009;1:RRN1134. https://doi.org/10.1371/currents.RRN1134.
122. Matrajt L, Longini IM. Optimizing vaccine allocation at different points in time during an epidemic. PLoS One. 2010;5(11):e13767. https://doi.org/10.1371/journal.pone.0013767.
123. Myliusa SD, Hagenaars TJ, Lugnera AK, et al. Optimal allocation of pandemic influenza vaccine depends on age, risk and timing. Vaccine. 2008;26(29-30):3742–9. https://doi.org/10.1016/j.vaccine.2008.04.043.
124. Bansal S, Pourbohloul B, Meyers LA. A comparative analysis of influenza vaccination programs. PLoS Med. 2006;3(10):e387. https://doi.org/10.1371/journal.pmed.0030387.
125. Yaesoubi R, Cohen T. Dynamic health policies for controlling the spread of emerging infections: influenza as an example. PLoS One. 2011;6(9):e24043. https://doi.org/10.1371/journal.pone.0024043.
126. Merl D, Johnson LR, Gramacy RB, et al. A statistical framework for the adaptive management of epidemiological interventions. PLoS One. 2009;4(6):e5807. https://doi.org/10.1371/journal.pone.0005807.
127. Chowell G, Viboud C, Wang X, et al. Adaptive vaccination strategies to mitigate pandemic influenza: Mexico as a case study. PLoS One. 2009;4(12):e8164. https://doi.org/10.1371/journal.pone.0008164.
128. National Vaccine Advisory Committee. Recommendations from the National Vaccine Advisory Committee: standards for adult immunization practice. Public Health Rep. 2014;129(2):115–23. https://doi.org/10.1177/003335491412900203.
129. US Department of Health and Human Services. HHS Pandemic Influenza Plan; 2005. Accessed 10 Nov 2005. Website, Available from: http://www.flu.gov/planning-preparedness/federal/hhspandemicinfluenzaplan.pdf.
130. Center for Health Protection. Human Swine Influenza Vaccination Programme Launched; 2009. Accessed 21 Dec 2009. Website, Available from: http://www.chp.gov.hk/en/content/116/19635.html.
131. Diekmann O, Heesterbeek JAP, Roberts MG. The construction of next-generation matrices for compartmental epidemic models. J R Soc Interface. 2010;7(47):873–85. https://doi.org/10.1098/rsif.2009.0386.
132. Casiday R, Cresswell T, Wilson D, et al. A survey of UK parental attitudes to the MMR vaccine and trust in medical authority. Vaccine. 2006;24(2):177–84. https://doi.org/10.1016/j.vaccine.2005.07.063.
133. Baker JP. The pertussis vaccine controversy in Great Britain, 1974-1986. Vaccine. 2003;21(25-26):4003–10. https://doi.org/10.1006/jmre.1997.1278.
134. Lau JT, Yeung NC, Choi K, et al. Acceptability of A/H1N1 vaccination during pandemic phase of influenza A/H1N1 in Hong Kong: Population based cross sectional survey. Br Med J. 2009;339:b4164. https://doi.org/10.1136/BMJ.b4164.
135. Latané B. The psychology of social impact. Am Psychol. 1981;36(4):343. https://doi.org/10.1037/0003-066X.36.4.343.
136. Latané B, Wolf S. The social impact of majorities and minorities. Psychol Rev. 1981;88(5):438. https://doi.org/10.1037/0033-295x.88.5.438.

137. Lau JTF, Yeung NCY, Choia KC, et al. Factors in association with acceptability of A/H1N1 vaccination during the influenza A/H1N1 pandemic phase in the Hong Kong general population. Vaccine. 2010;28(29):4632–7. https://doi.org/10.1016/j.vaccine.2010.04.076.
138. Zijtregtop EAM, Wilschut J, Koelma N, et al. Which factors are important in adults' uptake of a (pre)pandemic influenza vaccine? Vaccine. 2009;28(1):207–27. https://doi.org/10.1016/j.vaccine.2009.09.099.
139. Barriere J, Vanjak D, Kriegel I, et al. Acceptance of the 2009 A (H1N1) influenza vaccine among hospital workers in two French cancer centers. Vaccine. 2010;28(43):7030–4. https://doi.org/10.1016/j.vaccine.2010.08.021.
140. Rubin G, Potts H, Michie S. The impact of communications about swine flu (influenza A H1N1v) on public responses to the outbreak: Results from 36 national telephone surveys in the UK. Health Technol Assess. 2010;14(34):183–266. https://doi.org/10.3310/hta14340-03.
141. Maltezou HC, Dedoukou X, Patrinos S, et al. Determinants of intention to get vaccinated against novel (pandemic) influenza A H1N1 among health-care workers in a nationwide survey. J Infect. 2010;61(3):252–8. https://doi.org/10.1016/j.jinf.2010.06.004.
142. Hidiroglu S, Ay P, Topuzoglu A, et al. Resistance to vaccination: The attitudes and practices of primary healthcare workers confronting the H1N1 pandemic. Vaccine. 2010;28(51):8120–4. https://doi.org/10.1016/j.vaccine.2010.09.104.
143. Seale H, Heywood AE, McLaws ML, et al. Why do I need it? I am not at risk! Public perceptions towards the pandemic (H1N1) 2009 vaccine. BMC Infect Dis. 2010;10(1):99. https://doi.org/10.1186/1471-2334-10-99.
144. Wong LP, Sam IC. Factors influencing the uptake of 2009 H1N1 influenza vaccine in a multiethnic Asian population. Vaccine. 2010;28(28):4499–505. https://doi.org/10.1016/j.vaccine.2010.04.043.
145. Rubin GJ, Potts HW, Michie S. Likely uptake of swine and seasonal flu vaccines among healthcare workers. A cross-sectional analysis of UK telephone survey data. Vaccine. 2011;29(13):2421–8. https://doi.org/10.1016/j.vaccine.2011.01.035.
146. Torun SD, Torun F. Vaccination against pandemic influenza A/H1N1 among healthcare workers and reasons for refusing vaccination in Istanbul in last pandemic alert phase. Vaccine. 2010;28(35):5703–10. https://doi.org/10.1016/j.vaccine.2010.06.049.
147. Bauch CT, Earn DJD. Vaccination and the theory of games. Proc Natl Acad Sci. 2004;101(36):13391–4. https://doi.org/10.1073/pnas.0403823101.
148. Bauch CT. Imitation dynamics predict vaccinating behaviour. Proc R Soc B Biol Sci. 2005;272(1573):1669–75. https://doi.org/10.1098/rspb.2005.3153.
149. Reluga TC, Bauch CT, Galvani AP. Evolving public perceptions and stability in vaccine uptake. Math Biosci. 2006;204(2):185–98. https://doi.org/10.1016/j.mbs.2006.08.015.
150. Reluga TC, Galvani AP. A general approach for population games with application to vaccination. Math Biosci. 2011;230(2):67–78. https://doi.org/10.1016/j.mbs.2011.01.003.
151. Cojocaru MG. Dynamic equilibria of group vaccination strategies in a heterogeneous population. J Glob Optim. 2008;40(1-3):51–63. https://doi.org/10.1007/s10898-007-9204-7.
152. Perisic A, Bauch CT. Social contact networks and disease eradicability under voluntary vaccination. PLoS Comput Biol. 2009;5(2):e1000280. https://doi.org/10.1371/journal.pcbi.1000280.
153. Perisic A, Bauch CT. A simulation analysis to characterize the dynamics of vaccinating behaviour on contact networks. BMC Infect Dis. 2009;9(1):77. https://doi.org/10.1186/1471-2334-9-77.
154. Bauch CT, Bhattacharyya S. Evolutionary game theory and social learning can determine how vaccine scares unfold. PLoS Comput Biol. 2012;8(4):e1002452. https://doi.org/10.1371/journal.pcbi.1002452.
155. Fu F, Rosenbloom DI, Wang L, et al. Imitation dynamics of vaccination behaviour on social networks. Proc R Soc B Biol Sci. 2011;278(1702):42–9. https://doi.org/10.1098/rspb.2010.1107.
156. d'Onofrio A, Manfredi P, Poletti P. The impact of vaccine side effects on the natural history of immunization programmes: An imitation-game approach. J Theor Biol. 2011;273(1):63–71. https://doi.org/10.1016/j.jtbi.2010.12.029.

157. Mbah MLN, Liu J, Bauch CT, et al. The impact of imitation on vaccination behavior in social contact networks. PLoS Comput Biol. 2012;8(4):e1002469. https://doi.org/10.1371/journal.pcbi.1002469.
158. Zhang H, Zhang J, Li P, et al. Risk estimation of infectious diseases determines the effectiveness of the control strategy. Physica D. 2011;240(11):943–8. https://doi.org/10.1016/j.physd.2011.02.001.
159. Breban R, Vardavas R, Blower S. Mean-field analysis of an inductive reasoning game: Application to influenza vaccination. Phys Rev E. 2007;76(3):031127. https://doi.org/10.1103/PhysRevE.76.031127.
160. d'Onofrio A, Manfredi P, Salinelli E. Vaccinating behaviour, information, and the dynamics of SIR vaccine preventable diseases. Theor Popul Biol. 2007;71(3):301–17. https://doi.org/10.1016/j.tpb.2007.01.001.
161. Shim E, Meyers LA, Galvani AP. Optimal H1N1 vaccination strategies based on self-interest versus group interest. BMC Public Health. 2011;11(S1):S4. https://doi.org/10.1186/1471-2458-11-S1-S4.
162. Watts DJ, Dodds PS. Influentials, networks, and public opinion formation. J Consum Res. 2007;34(4):441–58. https://doi.org/10.1086/518527.
163. Nowak A, Szamrej J, Latané B. From private attitude to public opinion: A dynamic theory of social impact. Psychol Rev. 1990;97(3):362. https://doi.org/10.1037/0033-295X.97.3.362.
164. Salathé M, Bonhoeffer S. The effect of opinion clustering on disease outbreaks. J R Soc Interface. 2008;5(29):1505–8. https://doi.org/10.1098/rsif.2008.0271.
165. Bish A, Yardley L, Nicoll A, et al. Factors associated with uptake of vaccination against pandemic influenza: A systematic review. Vaccine. 2011;29(38):6472–84. https://doi.org/10.1016/j.vaccine.2011.06.107.
166. Young ME, Norman GR, Humphreys KR. Medicine in the popular press: The influence of the media on perceptions of disease. PLoS One. 2008;3(10):e3552. https://doi.org/10.1371/journal.pone.0003552.
167. Vance K, Howe W, Dellavalle RP. Social internet sites as a source of public health information. Dermatol Clin. 2009;27(2):133–6. https://doi.org/10.1016/j.det.2008.11.010.
168. Signorini A, Segre AM, Polgreen PM. The use of Twitter to track levels of disease activity and public concern in the US during the influenza A H1N1 pandemic. PLoS One. 2011;6(5):e19467. https://doi.org/10.1371/journal.pone.0019467.
169. Keelan J, Pavri-Garcia V, Tomlinson G, et al. YouTube as a source of information on immunization: a content analysis. J Am Med Assoc. 2007;298(21):2482–4. https://doi.org/10.1001/jama.298.21.2482.
170. Pandey A, Patni N, Singh M, et al. YouTube as a source of information on the H1N1 influenza pandemic. Am J Prev Med. 2010;38(3):e1–3. https://doi.org/10.1016/j.amepre.2009.11.007.
171. Shafer G. A Mathematical Theory of Evidence. vol. 76. London: Princeton University Press; 1976.
172. Dempster AP. Upper and lower probabilities induced by a multivalued mapping. In: Classic Works of the Dempster-Shafer Theory of Belief Functions. Berlin: Springer; 2008. p. 57–72.
173. Information Services Department of the Hong Kong Government. Human Swine Influenza Vaccination Programme; 2010. Accessed 1 April 2011. Website, Available from: http://www.info.gov.hk/gia/general/201003/17/P201003170168.htm.
174. Panzarasa P, Opsahl T, Carley KM. Patterns and dynamics of users' behavior and interaction: network analysis of an online community. J Am Soc Inf Sci Technol. 2009;60(5):911–32. https://doi.org/10.1002/asi.v60:5.
175. Funk S, Gilada E, Watkinsb C, et al. The spread of awareness and its impact on epidemic outbreaks. Proc Natl Acad Sci. 2009;106(16):6872–6877. https://doi.org/10.1073/pnas.0810762106.
176. Fenichel EP, Castillo-Chavez C, Ceddia MG, et al. Adaptive human behavior in epidemiological models. Proc Natl Acad Sci. 2011;108(15):6306–11. https://doi.org/10.1073/pnas.1011250108.

177. Coelho FC, Codeco CT. Dynamic modeling of vaccinating behavior as a function of individual beliefs. PLoS Comput Biol. 2009;5(7):e1000425. https://doi.org/10.1371/journal.pcbi.1000425.
178. Kaplan AM, Haenlein M. Users of the world, unite! The challenges and opportunities of social media. Bus Horiz. 2010;53(1):59–68. https://doi.org/10.1016/j.bushor.2009.09.003.
179. Ginsberg J, Mohebbi MH, Patel RS, et al. Detecting influenza epidemics using search engine query data. Nature. 2009;457(7232):1012–4. https://doi.org/10.1038/nature07634.
180. Salathe M, Khandelwal S. Assessing vaccination sentiments with online social media: Implications for infectious disease dynamics and control. PLoS Comput Biol. 2011;7(10):e1002199. https://doi.org/10.1371/journal.pcbi.1002199.
181. Henrich N, Holmes B. What the public was saying about the H1N1 vaccine: Perceptions and issues discussed in on-line comments during the 2009 H1N1 pandemic. PLoS One. 2011;6(4):e18479. https://doi.org/10.1371/journal.pone.0018479.
182. Xia S, Zhou XN, Liu J. Systems thinking in combating infectious diseases. Infect Dis Poverty. 2017;6(1):144. https://doi.org/10.1186/s40249-017-0339-6.
183. Checkland P. Systems thinking, systems practice: Includes a 30-year retrospective. J Oper Res Soc. 2000;51(5):647. https://doi.org/10.2307/254200.
184. Leischow SJ, Milstein B. Systems thinking and modeling for public health practice. Am Public Health Assoc. 2006;96(3):403–5. https://doi.org/10.2105/ajph.2005.082842.
185. Maani K, Cavana RY. Systems Thinking, System Dynamics: Managing Change and Complexity. New Zealand: Pearson Education New Zealand; 2007.
186. May RM. Simple mathematical models with very complicated dynamics. Nature. 1976;261(5560):459–67. https://doi.org/10.1038/261459a0.
187. Liu J, Jin X, Tsui KC. Autonomy Oriented Computing: From Problem Solving to Complex Systems Modeling. Boston: Springer; 2006.
188. Liu J. Autonomous Agents and Multi-agent Systems: Explorations in Learning, Self-organization and Adaptive Computation. Singapore: World Scientific; 2001.
189. Newman ME. Spread of epidemic disease on networks. Phys Rev E. 2002;66(1):016128. https://doi.org/10.1103/PhysRevE.66.016128.
190. Chen H, Yang B, Liu J, et al. Mining spatiotemporal diffusion network: A new framework of active surveillance planning. IEEE Access. 2019;7:108458–73. https://doi.org/10.1109/ACCESS.2019.2927878.
191. Yang X, Liu J, Zhou XN, et al. Inferring disease transmission networks at a metapopulation level. Malar J. 2014;2(1):8. https://doi.org/10.1186/2047-2501-2-8.
192. Shi B, Zhong J, Bao Q, et al. EpiRep: Learning node representations through epidemic dynamics on networks. In: 2019 IEEE/WIC/ACM International Conference on Web Intelligence. Piscataway: IEEE; 2019. p. 486–92.
193. Yu Z, Liu J, Wang X, et al. Efficient vaccine distribution based on a hybrid compartmental model. PLoS One. 2016;11(5):e0155416. https://doi.org/10.1371/journal.pone.0155416.
194. Liu J, Yang B, Cheung WK, et al. Malaria transmission modelling: A network perspective. Infect Dis Poverty. 2012;1(1):11. https://doi.org/10.1186/2049-9957-1-11.
195. Ming RX, Liu J, Cheung WK, et al. Stochastic modelling of infectious diseases for heterogeneous populations. Infect Dis Poverty. 2016;5(1):107. https://doi.org/10.1186/s40249-016-0199-5.
196. Yang B, Pei H, Chen H, et al. Characterizing and discovering spatiotemporal social contact patterns for healthcare. IEEE Trans Pattern Anal Mach Intell. 2016;39(8):1532–46. https://doi.org/10.1109/TPAMI.2016.2605095.
197. Chen H, Yang B, Pei H, et al. Next generation technology for epidemic prevention and control: Data-driven contact tracking. IEEE Access. 2018;7:2633–42. https://doi.org/10.1109/ACCESS.2018.2882915.
198. Shi B, Tan Q, Zhou XN, et al. Mining geographic variations of Plasmodium vivax for active surveillance: A case study in China. Malar J. 2015;14(1):216. https://doi.org/10.1186/s12936-015-0719-y.

199. Shi B, Liu J, Zhou XN, et al. Inferring Plasmodium vivax transmission networks from tempo-spatial surveillance data. PLoS Negl Trop Dis. 2014;8(2):e2682. https://doi.org/10.1371/journal.pntd.0002682.
200. Shi B, Xia S, Yang GJ, et al. Inferring the potential risks of H7N9 infection by spatiotemporally characterizing bird migration and poultry distribution in eastern China. Infect Dis Poverty. 2013;2(1):8. https://doi.org/10.1186/2049-9957-2-8.
201. Zhu G, Liu J, Tan Q, Shi B. Inferring the spatio-temporal patterns of dengue transmission from surveillance data in Guangzhou, China. PLoS Negl Trop Dis. 2016;10(4):e0004633. https://doi.org/10.1371/journal.pntd.0004633.
202. Shi B, Zhan XM, Zheng JX, et al. Identifying key bird species and geographical hotspots of avian influenza A (H7N9) virus in China. Infect Dis Poverty. 2018;7(1):97. https://doi.org/10.1186/s40249-018-0480-x.
203. Zhang Y, Cheung WK, Liu J. A unified framework for epidemic prediction based on poisson regression. IEEE Trans Knowl Data Eng. 2015;27(11):2878–92. https://doi.org/10.1109/TKDE.2015.2436918.
204. Pei H, Yang B, Liu J, et al. Group sparse Bayesian learning for active surveillance on epidemic dynamics. In: 32nd AAAI Conference on Artificial Intelligence. Menlo Park: AAAI; 2018. p. 800–7.
205. Chen H, Yang B, Liu J. Partially observable reinforcement learning for sustainable active surveillance. In: International Conference on Knowledge Science, Engineering and Management. Berlin: Springer; 2018. p. 425–37.
206. Yang B, Guo H, Yang Y, et al. Modeling and mining spatiotemporal patterns of infection risk from heterogeneous data for active surveillance planning. In: 28th AAAI Conference on Artificial Intelligence. Menlo Park: AAAI; 2014. p. 493–9.
207. Shi B, Zheng J, Qiu H, et al. Risk assessment of malaria transmission at the border area of China and Myanmar. Infect Dis Poverty. 2017;6(1):108. https://doi.org/10.1186/s40249-017-0322-2.
208. Tan Q, Liu J, Shi B, et al. Public health surveillance with incomplete data-spatio-temporal imputation for inferring infectious disease dynamics. In: 6th IEEE International Conference on Healthcare Informatics. Piscataway: IEEE; 2018. p. 255–64.
209. Shi B, Xia S, Liu J. A complex systems approach to infectious disease surveillance and response. In: International Conference on Brain and Health Informatics. Berlin: Springer; 2013. p. 524–35.

Index

J. Liu, S. Xia, *Computational Epidemiology*, Health Information Science,
https://doi.org/10.1007/978-3-030-52109-7

V

W

Z

The manufacturer's authorised representative in the EU is Springer Nature Customer Service Centre GmbH, Europaplatz 3, 69115 Heidelberg, Germany. If you have any concerns regarding our products, please contact ProductSafety@springernature.com

Printed and bound by CPI Group (UK) Ltd, Croydon, CR0 4YY
22/09/2026
02221060-0018